DSP-Based Testing of Analog and Mixed-Signal Circuits

Matthew Mahoney

IEEE Computer Society Press
Los Alamitos, California

Washington • Brussels • Tokyo

IEEE Computer Society Press
10662 Los Vaqueros Circle
P.O. Box 3014
Los Alamitos, CA 90720-1314

IEEE Computer Society Press Order Number BP00785
Library of Congress Number 87-80432
ISBN 0-8186-0785-8

Additional copies can be ordered from

IEEE Computer Society Press
Customer Service Center
10662 Los Vaqueros Circle
P.O. Box 3014
Los Alamitos, CA 90720-1314
Tel: (714) 821-8380
Fax: (714) 821-4641
Email: cs.books@computer.org

IEEE Service Center
445 Hoes Lane
P.O. Box 1331
Piscataway, NJ 08855-1331
Tel: (908) 981-1393
Fax: (908) 981-9667
mis.custserv@computer.org

IEEE Computer Society
13, avenue de l'Aquilon
B-1200 Brussels
BELGIUM
Tel: +32-2-770-2198
Fax: +32-2-770-8505
euro.ofc@computer.org

IEEE Computer Society
Ooshima Building
2-19-1 Minami-Aoyama
Minato-ku, Tokyo 107
JAPAN
Tel: +81-3-3408-3118
Fax: +81-3-3408-3553
tokyo.ofc@computer.org

The Institute of Electrical and Electronics Engineers, Inc.

Preface

There are many textbooks devoted to digital signal processing (DSP), but very few that relate DSP to the special concerns of automated testing. Those aimed at graduate study focus on theory, derivations, and proofs, while applications-oriented texts tend to deal with problems of communications engineering or laboratory research.

Knowing DSP theory is not the same as knowing what to do when asked to test a million automobile radios, or flash converters, or compact disc D/A converters, or CODECs, or video amplifiers, or switched-capacitor filters. DSP theory alone does not tell you how to measure nanovolts per root hertz in a production environment, or 0.01 degrees of phase shift, or -110 dB ninth harmonic distortion.

Neither does a knowledge of electronic engineering, by itself. Nor does traditional experience with automatic test equipment (ATE). It is the union of ATE and DSP that has proved most fruitful in the last few years, providing effective solutions to increasingly difficult test problems: those demanding laboratory accuracy at production speed.

The papers in this tutorial specifically address such problems and solutions. In writing and assembling this material, it was my intent to provide a bridge between DSP theory and the physical world of electrical components and circuits. Since there are a number of excellent texts that develop the underlying principles, I have tried to emphasize instead the mathematical concepts rather than the derivations—to show the forest instead of the trees, so to speak. As much as anything, my goal was to establish a philosophy of DSP-based testing: How to think, how to approach a problem, how to create a solution, and how to determine if it really works properly. My hope is that even the reader not directly involved with testing will find food for thought in the concepts.

There are three sections to the text. The first section, divided into 12 chapters, is drawn largely from material originally prepared as class notes and seminar handouts, but which have been updated and expanded to stand alone, without the instructor's presence. The second part (Chapter 13) consists of reprints of five papers presented at past IEEE International Test Conferences, and which supplement the techniques covered in Chapters 1 through 12. The last section begins with a short list of current references, and concludes with a large and historically significant bibliography reprinted from a highly recommended IEEE Press reference [8], *Digital Signal Processing* by Rabiner and Gold.

I would like to thank my colleagues at LTX Corporation for their support and assistance. I also want to thank Ken Anderson, Test Technology Technical Committee of the Computer Society of the IEEE, for convincing me to undertake this project, and Margaret Brown and the publications staff (especially Lee Blue and Denise Felix), Computer Society of the IEEE, for their patience and hard work in bringing this tutorial text to completion.

Matthew Mahoney

Table of Contents

Chapter 1
Introduction to DSP-Based Testing

Chapter 1: Introduction to DSP-Based Testing

In the last few years, digital signal processing (DSP) has profoundly altered the design and use of automatic test equipment (ATE). One of the most significant changes is that the ATE computer, instead of simply controlling and monitoring hardware instruments, can now emulate and replace them.

In this tutorial, test systems that use the computer as a substitute for instruments are termed DSP-based machines. Ideally, such systems contain no conventional analog instruments whatsoever. The only electronic circuits are those of the computer, the peripheral devices, power supplies, and interface circuits to the device under test (DUT).

It might seem that these machines, which have no analog test circuits, are aimed at digital testing, but that is not the case. Only the physical bodies of the analog instruments are missing, not their functions. The instruments are still there, in the form of computer models.

Substituting software routines for physical circuits provides an effective way around many otherwise unavoidable limits of analog instrumentation: crosstalk, nonlinearity, noise, drift, aging, improper calibration, filter settling time, thermal effects, and so on. Thus, while DSP can indeed perform purely digital test functions, its primary commercial appeal lies in the improvements it makes in testing complex analog and mixed-signal (A/D/A) devices. For manufacturing, the fact that emulated circuits operate faster than their analog counterparts means higher test throughput. For engineering, the fact that they eliminate many analog errors means improved repeatability and accuracy. For incoming inspection, the ability to connect, adjust, and even create instruments from the keyboard means vastly increased test flexibility.

How do these machines work? What role does DSP play in the process? The articles that follow were written specifically to answer these questions and to show how DSP performs in real test situations.

Overview of Testing

DSP-based test equipment differs substantially from what is commonly called ATE. To understand this difference, it helps to review not only the structure of conventional ATE but also the concepts which underly the general practice of testing.

The terms, *test* and *measurement*, are frequently used interchangeably, but they really describe different processes. *Measurement* is a process of quantification (i.e., of obtaining a descriptive numerical value for some property or phenomenon). Although judgment may follow, that is a separate consideration. A good part of laboratory measurements are aimed only at learning how things behave, not whether they "pass" or "fail."

Testing, by contrast, is a process of grading and sorting things, to determine their acceptability for a given application. It most often involves the application of a *stimulus* and a judgment of the *response*.

In this tutorial, the objects or "devices," are semiconductor circuits and subsystems, especially complex analog circuits and mixed-signal circuits. But the definition extends to almost anything. If you manufactured coil springs, for example, you might want to test each one for its deflection rate; if the stimulus is an applied "reference" force, the spring should respond by deflecting a certain distance, within specified limits. If there are different limits for different grades of springs, they would be sorted into different *bins* or boxes accordingly. This concept of "binning" is retained in ATE software today, but is more likely to direct the output chute of an IC handler.

While measurement is definitely involved in most analog and mixed-signal (analog-digital) tests, it is not required in all testing. We can grade devices by comparison, for example, without ever determining numerical values. In fact, a test usually goes much faster if measurement is not required. Procedures of this type are often called *go/no go* tests with physical objects, or *pass/fail tests* with electrical components.

Digital testing provides an example of real-time, pass/fail testing. The principle is outlined in Figure 1.1, where a pattern of 1s and 0s is applied to the DUT. Since a digital circuit is a *deterministic* device (having a completely definable set of input-output states), it can be tested by comparing its logic output pattern against a precomputed "compare" pattern, cycle-by-cycle. If any bits fail to match, the device fails.

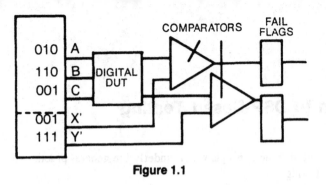

Figure 1.1

But this is getting ahead of ourselves. The conventional, or pre-DSP, approach to analog or "linear" circuit testing stems from bench set-ups in the 1930s and 1940s when the typical test hardware or *fixture* consisted of a DC source or sinewave generator as the input stimulus, plus a meter or oscilloscope to measure the response. Generally, one or more filters were employed to allow selective waveform examination.

Figure 1.2 illustrates the traditional approach to *transmission testing*, involving two informational "ports." The stimulus is applied at one port and the response observed at the other port.*

DSP is of no direct value in this kind of testing because there is no "signal" to be processed or analyzed. (DSP algorithms can, however, help to prepare the digital drive and compare patterns in advance.)

The real value of DSP in testing, and in this tutorial, is in evaluating *non-deterministic devices*, namely, analog and mixed signal circuits. As with coil springs, no two analog DUTs will respond identically. In contrast to digital testing, however, error is permissible as long as it stays within specified limits. In most electrical tests, such error cannot be properly analyzed until the whole output waveform or sequence is processed as an entity, and this is where DSP is most valuable.

Automation did not change the basic form of Figure 1.2. As "linear" ATE evolved during the 1970s, the minicomputer was viewed mostly as a controller and data logger (i.e., as a means of replacing the human operator). Analog source and measurement instruments were still present, operating in much the same way. Just as in the manual fixture, there was no frequency-time synchronization between the left and right halves of the fixture, in contrast to the digital fixture of Figure 1.1.

---

*Commercial testing also involves so-called *parametric testing*, in which the stimulus and response are at a single port. In parametric testing, the measured property is usually a simple one like leakage current, output impedance, or capacitance.

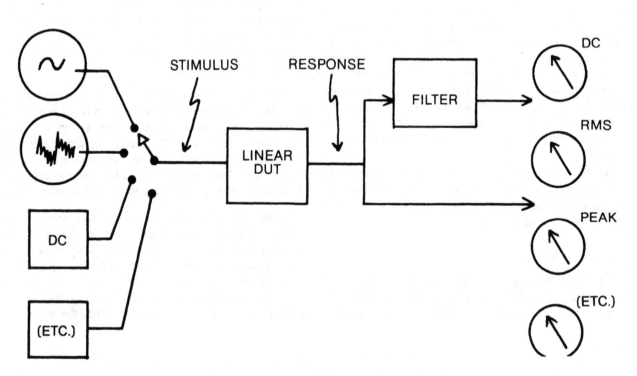

Figure 1.2: Traditional concept of analog transmission testing

Emulation versus Automation

DSP systems represent a distinct departure from the preceding concept of ATE. In fact, it is probably incorrect to refer to a DSP test system as "automated" test equipment.

Automation generally implies a process in which equipment is modified or otherwise adapted to operation by an electrical or mechanical controller instead of a human. Generally, this modification does not alter the basic principle by which the equipment performs its end task. A numerically controlled lathe, for example, removes metal by the same fundamental cutting process as a manually controlled metal lathe.

This concept of automation is the underlying concept of classical linear ATE. The idea is to provide hardware "resources"—the collective name for various instrumentation modules—that electrically function in much the same way as the circuits of good bench instruments. Rather than being operated by hand and eye, however, these resources are switched, adjusted, and monitored by minicomputer far faster than by human hand and eye. Figure 1.3 illustrates the resource concept.

Invisible Instruments

In DSP-based testing, by contrast, the computer does not automate the resources; it *replaces* them. This is an important distinction, because it is easy to conclude from the name alone that a DSP test system is simply a conventional system with added DSP software.

Another mistaken impression is that a DSP machine evaluates the test device by new and different parameters or that it deals with statistical properties rather than electrical ones. In early trials of DSP testing, this was partly true and, in fact, may have delayed acceptance of the principle.

Ultimately, however, the success or failure of any new test scheme depends on how well it correlates with accepted standards. Thus, while a DSP system can indeed provide statistical analysis, this ability is intended to supplement, not replace, the task of measuring familiar electrical parameters. A good DSP tester should be able to *emulate* (electrically imitate) a wide range of existing instruments and to exercise the DUT with accepted types of signals.

To do this, a DSP tester needs a computer programmed to simulate the functions of conventional ATE resources. There must also be special interface circuits that enable the computer to communicate with the DUT. For both speed and convenience, the simulation should not be done by high-level routines but should be built into the computer's software operating system. These transparent routines are the DSP machine's equivalent of hardware resources and serve as

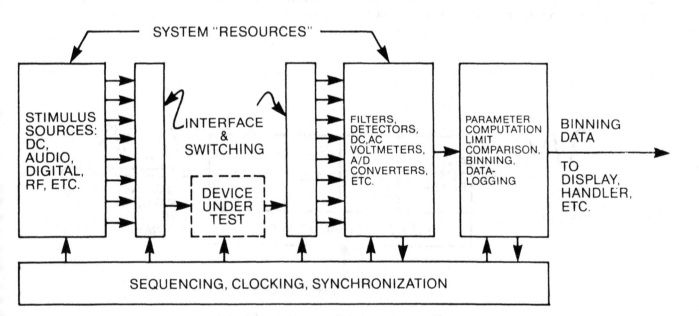

Figure 1.3

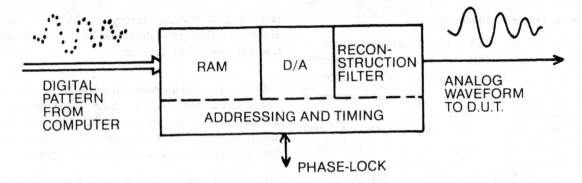

Figure 1.4: Concept of waveform synthesizer

modular instruments from which a variety of fixtures can be assembled under keyboard control.

Numerical Vectors

Instead of actually measuring voltages or currents, DSP instruments in reality process *numerical vectors*. These are strings of informationally related numbers that represent sampled waveforms, spectra, filter responses, or any function for which a curve could be drawn. *A DSP-based system is one that allows you to label, create, transfer, and analyze numerical vectors as simply as you would manipulate single variables in a hand calculator.*

To create a stimulus signal in such a system, the test engineer defines the stimulus and (with the computer's help) converts this description to a vector; that is, to a string of data points that traces out the desired waveshape. The vec-

tor is expressed in integers and is transferred as a single entity (a vector transfer) to the local memory of a *waveform synthesizer* (Figure 1.4). This pattern is fed to a digital-to-analog converter (DAC), usually in a continuous loop. If a transient waveform is desired, the loop is terminated.

The DAC output is "de-glitched" (a type of "hold" function) to remove transition imperfections, then passed through a *reconstruction* filter to obtain continuous, band-limited waveforms. In tests that call for stepwise waveforms, the filter is bypassed.

The process is reversed at the DUT output. The analog response waveform is converted into numerical (vector) form by a *waveform digitizer*, and sent to the DSP "instruments" for analysis (Figure 1.5).

The procedure just described assumes that the DUT is an all-analog device. For devices that produce digital output responses (e.g., A/D converters (ADCs)), the output pat-

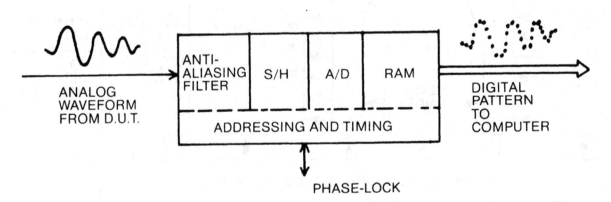

Figure 1.5: Concept of waveform digitizer

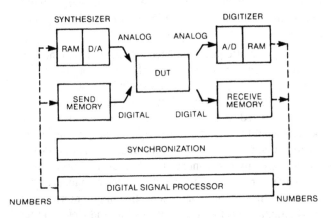

SYNTHESIZER   DIGITIZER

Figure 1.6: Fundamental structure of DSP-based test system

tern is collected by a temporary RAM called the *receive memory* and then sent as a vector to the processor. Similarly, if the DUT input is digital instead of analog, the stimulus vector would not be sent to the synthesizer, but to a *transmit* or *send memory* and "replayed" to the DUT input under local timing control.

The basic form of a mixed-technology DSP-based test system is shown in Figure 1.6 and comprises seven key functions. These are the four RAM-based units listed above, plus three more functions that provide coordination for the system: high-speed (vector) buses, time-lock or phase-lock synchronization, and one or more computers especially equipped for vector processing.

Vector Transfer

In conventional ATE, data are commonly transferred one word at a time, alternating with address information. For DSP systems, it is more efficient to transmit related data as contiguous blocks of words (i.e., vectors), accompanied only by starting address and vector length.

Systems that transfer data this way are said to have a *burst* I/0, or DSP bus or *vector* bus architecture. Ideally, such buses are separate from the system data and control buses and directly connect the memories of the various RAMs and

processors. The intent is to keep vector transfer time small in comparison with other aspects of test time. A transfer time around 1 millisecond (ms) for 1 K 16-bit integers is typical. By comparison, it takes about 5 ms to connect or disconnect a test circuit by relay switching.

Vector and Array Processing Speed

Consider an audio amplifier to be tested for gain and distortion at 1000 Hz. If all distortion is harmonically related to the fundamental, then all the information needed to analyze the various components occurs within 1 cycle. However, analysis cannot start until the device output has settled to its steady state AC condition, perhaps 1 ms or less for a typical audio amplifier.

In principle, an infinitely fast test system could perform the required tests in about 2 ms. This is the so-called "intrinsic" test time: the limit imposed by the physics of the device and by the nature of the test.

How do practical systems compare? Whether DSP or conventional ATE, most test plans begin with the same procedure:

1. Connect an audio source or synthesizer and set it to the proper frequency and amplitude.

2. Connect an audio voltmeter or digitizer to the DUT output, and set it to the appropriate range.

3. Wait for all circuits to settle to steady state output.

The next step depends upon the type of test system. With DSP, the waveform is digitized and is then sent back to the computer for analysis. With analog instrumentation, analysis is performed by a detector circuit that produces a DC voltage proportional to the parameter magnitude. This voltage is then sent to an analog-to-digital converter (ADC).

In the DSP scheme, the ADC precedes the operation of analysis and output is a vector. In analog ATE, the ADC follows the operation of analysis and its output is a scalar (i.e., a single sample). In either arrangement the computer compares the answers against predetermined limits and acts accordingly.

Even with modern equipment, the collective time required for these tasks is often quite large when compared with intrinsic test time. "Rack-and-stack" systems built from bus-operated bench instruments may take a second or more for each different frequency component. Integrated analog systems can do much better but are still far from perfect. The following figures are typical of automated 1 kHz amplitude measurement:

Relay switching	5 ms
Filter + detector settling	35 ms (includes source and DUT)
Computer overhead	10 ms
Total time	50 ms

To measure the second harmonic, the 1 kHz filter would be replaced by a 2 kHz bandpass filter, and the process repeated. The total time thus depends on the number of frequency components to be measured.

In DSP-based testing, the combined operation of filtering and measurement can be provided by the discrete Fourier transform (DFT) or by the fast fourier transform (FFT). The DFT is useful in analyzing a single spectral line, whereas the FFT is best for obtaining many lines at once. (Using the FFT is faster than repeating the DFT many times.) These operations, and most others that involve sums-of-products expressions, are performed at high speed by a special auxiliary computer called an *array processor.* Commercial board-level array processors can provide a full spectral analysis of a typical vector via FFT in roughly 4 to 20 ms.

Using the FFT, a representative array processor-equipped DSP system might take about 30 ms to perform the previous 1 kHz gain and distortion test:

Relay switching	5 ms
Load and start synthesizer	5 ms
Synthesizer/DUT settling	1 ms
Digitization interval	1 ms (minimum)
Transfer time	1 ms
Processing time + overhead	15 ms
Total time	28 ms (minimum)

In later chapters, we will see that it is necessary to capture many signal cycles for certain measurements. For simple analog tests, however, a few cycles will suffice, and 30 ms is a reasonable total.

Although faster than the analog approach, DSP does not offer enough speed improvement for this one test to be economically significant. In fact, if gain were the *only* thing to be measured, an analog tester might be the wiser choice.

The real speed advantage of DSP becomes apparent when the DUT is to be tested for many parameters. First, relay switching occurs only once, since the synthesizer and

digitizer are left in place throughout the various tests. Second, all components related to a common stimulus are captured in a single waveform vector (i.e., the device need be exercised only once for each unique stimulus condition). Third, the mathematics is done independently of device status and may run concurrently with device handling, relay switching, digitizing, etc. DSP thus increases the *throughput,* or number of tests per unit time.

In the procedure just completed, for example, all the spectral components are computed by the FFT, not just the fundamental component. With only a slight increase in test time, we could examine dozens of harmonics and learn the relative phase. If 10 parameters were evaluated, the *average* time per test would be about 3.5 ms, versus nearly 50 ms for the analog system. Beyond that is the added attraction that we can analyze any component of the signal, not merely those for which a hardware filter and/or detector happens to be available. In addition, components that might interfere with conventional detectors (e.g., power line and ripple components) can be easily identified, measured, and set to zero before other measurements are made.

Processor Speed

Since the speed of DSP testing depends heavily on the speed of the processor, it should be pointed out that minicomputers, even those with mathematical accelerators or coprocessors, are often too slow at *vector* mathematics to be useful in commercial testing. To appreciate this, consider that a VAX 11/780 equipped with floating-point (FP) accelerator takes roughly 400 ms to perform a 1024-point "library" FFT routine. Personal computers are even slower. In a speed comparison done for this tutorial, an IBM PC/AT equipped with an 80287 coprocessor took more than 3 seconds to execute a similiar FFT routine.

These numbers are good when compared with nonaccelerated minicomputers, but are obviously too slow to replace analog test hardware. The problem is twofold: One is that the routines were written in a high-level language (FORTRAN), and the other is that conventional accelerators and coprocessors are designed to handle *scalar* mathematics, operations that involve only 1 or 2 input operands at a time and produce 1 output operand. To perform just 1 vector operation, scalar computers have to perform thousands of FP computations serially.

Vector mathematics calls for a different computational architecture, ideally one with all parallel computation. At present, the most cost-effective structure is that of the array processor, a special-purpose computer designed to process subsets of the vector elements in parallel and move the intermediate results along a "pipeline," or mathematical assembly line. This structure also reduces the number of store-and-fetch operations when compared with scalar architecture. Pipelines are slow for scalar operations, however, so array processors are primarily designed for vector and matrix mathematics.

The array processor gains additional speed because its routines are executed at machine level, often by dedicated logic. To an extent, this can (and should) be done in low-cost DSP systems that do not have an array processor. In many cases, simply building the DSP algorithms into the operating system of a good scalar computer provides enough DSP speed to satisfy low-volume test needs.

Floating-Point Mathematics

Speed is not the only requirement for vector processing, of course; accuracy is equally important. In most cases, anything less than 32-bit, FP mathematics will noticeably restrict the performance of the emulated instruments.

Why should this be so? If vectors originate in, or terminate in, 12- to 16-bit ADCs and DACs, why isn't 16-bit fixed-point mathematics sufficient? The motivation in asking this question is the possibility of implementing a small, low-cost DSP tester with the inexpensive 16-bit fixed-point chips readily available today. For a number of reasons, however, processing resolution has to be far greater than the resolution of the individual samples in a vector.

One difference is that the converter does not combine or manipulate samples, whereas the processor does. Consider two digitized samples from a 16-bit ADC, each with 8 leading zeros and 8 active bits. This might be more than enough resolution for a useful answer. But suppose the algorithm called for the product of these two samples. In fixed-point 16-bit format, if each number were treated as a fraction, the product would have 16 leading zeros. The information contained in the two samples would be lost forever!

Integer representation does not help but simply moves the trouble to the other end of the word. Multiplying one 16-bit integer by another merely forces the most significant bits out of a fixed-point 16-bit result.

This phenomenon is sometimes called the "black hole" effect and is a hazard of fixed-point multiplication and squaring. A partial solution is to prescale fixed-point operands so that (in the example of fractional representation) the largest number to be multiplied has no leading zero bits. This technique is not uncommon in low-cost array processors and DSP chips and is sometimes referred to as "block" FP format.

To process a vector, or "block," of 1024 elements, a block FP processor examines all 1024 elements before beginning the computation, and then shifts all elements by an equal number of binary places, so as to remove the leading zeros from the largest word.

This lessens the black hole effect, but does not not eliminate it, since the majority of block elements will still have leading zeros. Moreover, scaling takes time, sometimes more than the mathematical operation itself.

A better solution, and one that will be assumed throughout this tutorial, is the use of true FP hardware. This avoids all leading zeros in fractions and eliminates the black hole effect. It is fast because the scale factor is carried along with each word. To get 16 bits of resolution within a FP word requires more than 16 bits, of course. With 8 bits for sign and exponent, and 16 "mantissa" bits, for example, word length expands to 24 bits.

The closest commercial DSP format is 32-bit FP, with most variations having 22 to 24 mantissa bits. At first glance, this may seem to be more bits than is needed to solve the original problem. As it turns out, however, this extra resolution can be put to good use and, in fact, may be insufficient for certain computations.

First, vectors contain many samples, and the signal-to-quantization noise of the entire vector can be better than that of a single sample by (as much as) the square root of N. Given 1024 uniformly distributed samples over a prime number of signal cycles, the quantization noise appearing in any one spectral location will be reduced by the factor 32, or nearly 30 decibels. This is equivalent to 5 additional bits of resolution. We will analyze this in a later chapter, but it can be seen right away that a 21-bit mantissa is desirable to take full advantage of 16-bit digitizers.

Even greater mathematical resolution is desirable because many algorithms produce cumulative error. One example is an iterative procedure that computes angles by successive addition, using a modulus of 360 degrees. The angle never grows over 360, but an error in the initial or "seed" angle continues to grow with each addition. To allow all such algorithms to be used, one rule of thumb states that the mathematical processor should have at least three decimal orders of precision beyond what is desired in the end result. This translates to about 10 bits more than the 21 or so already established and suggests that a sufficiently precise DSP system needs more *internal* computational precision than even the standard 32-bit floating-point (FP) format provides. Modern array processors meet this by using extended precision in internal computation (e.g., 40 bits for the Texas

Instruments TMS320C30 chip) or by double-precision FP mathematics (64 bits).

Phase-Lock Synchronization

Of course, the foregoing mathematical precision will be wasted unless the vector samples fall in exactly the right places over exactly the right time interval. In good DSP test systems, the digitizing window must be precisely coordinated with each and every clock, signal, and distortion component. These will be discussed in some detail in later chapters. Here, it is sufficient just to point out that "synchronization" in Figure 1.6 means far more than simply clocking everything together. This final section of the 7-element structure usually involves a variety of frequency dividers and/or special circuits called phase-locked loops (PLLs). These produce uniformly distributed clock pulses and precisely timed windows, such that all rates and times can be programmed in integer ratios, often involving prime numbers.

Synchronization of this particular kind goes by various names, including M/N synchronization, integer-ratio synchronization, or prime-ratio locking. A system, in which all frequency and time functions are programmably related in exactly whole-number ratios, is said to be *coherent*.

Precise, repeatable, and programmable control of timing is taken for granted in digital circuit testing, but is almost unknown in classical linear circuit testing. You can see, however, that restricting analysis to a whole number of cycles is essential to accuracy, and the ability to select an arbitrary number enables the programmer to establish the best trade-off between speed and accuracy. It also provides highly repeatable results, and makes phase and delay measurements practical in production. We will explore these as well as other, not so obvious, benefits in later chapters.

Representative Digitizer

Commercial digitizers and synthesizers are more complex than the earlier sketches suggest. Figures 1.7 and 1.8 show the block diagrams of two representative units made by the LTX Corporation: the audio-range WS800 Waveform Synthesizer and the companion WD800 Waveform Digitizer, which are board-level units that are part of a large modular system, and have sampling rates to over 100 ks/s (kilosamples per second).

The synthesizer uses a 16-bit DAC to produce stepwise patterns. In certain tests, these patterns are used directly, while in others they are passed through a programmable reconstruction filter to produce continuous, band-limited analog waveforms. The filters are flat-topped multiple low-pass

units. Sine-X-over-X correction, where appropriate, is applied to the spectrum of the signal during digital synthesis of the pattern, and is thus built into the vector. (This is discussed in Chapter 3.)

The vector is stored in the synthesizer's local memory, a 16 K by 16-bit RAM. This can be subdivided into as many as 128 zones and enables on-the-fly switching from one waveshape to another. One use of this feature is in synthesizing phase-shift-keyed (PSK) communications signals.

The pattern may be clocked by an external source but is most often taken from one of the two post dividers (LI and L2) following the PLL. The clock may be switched from 1 divider output to the other under external bit control to produce phase-continuous frequency-shift keyed (FSK) stimulus signals. The divider input may also be taken from a continuously variable oscillator for "warping" (i.e., to produce conventional, continuous frequency modulation (FM)).

The synthesized analog waveform is sent through programmable coarse and fine attenuators to set the proper level and is then split into two paths: one, through a 50 ohm buffer, and the other, through a Hi-Z (600 ohms or more) buffer. Programmable DC offset may be added to either output. The result signal(s) may be sent to the test device through dedicated lines, or through a test head matrix. It may also be sent to the system's master DC voltmeter for step-by-step calibration.

Figure 1.8 shows the companion audio-range digitizer, which employs a linear 15-bit 100 ks/s ADC. The digitizer input path contains a differential buffer, a programmable-gain amplifier (PGA), and a programmable anti-aliasing filter. The ADC contains a built-in track-hold circuit. Digitized samples are temporarily stored in local bit RAM, until the desired vector is collected, and then transferred to the system computer and/or array processor. The waveform to be digitized may be taken from the DUT via dedicated lines or via the test head matrix. It may also be obtained from a system DC reference for sample-by-sample calibration if desired.

To permit this unit to be used with synthesizers and signal sources that do not contain their own PLLs, three independent PLLs are provided. In telecom testing, there is often the need to generate several clock rates that are different but precisely coordinated.

PLL 3 has several independent output dividers to permit coherent clocking of devices at different rates from this single PLL. Suppose, for example, that it were necessary to clock the digitizer at 50 ks/s, but lock a sine wave generator at 257/256 times this rate. The voltage-controlled oscillator

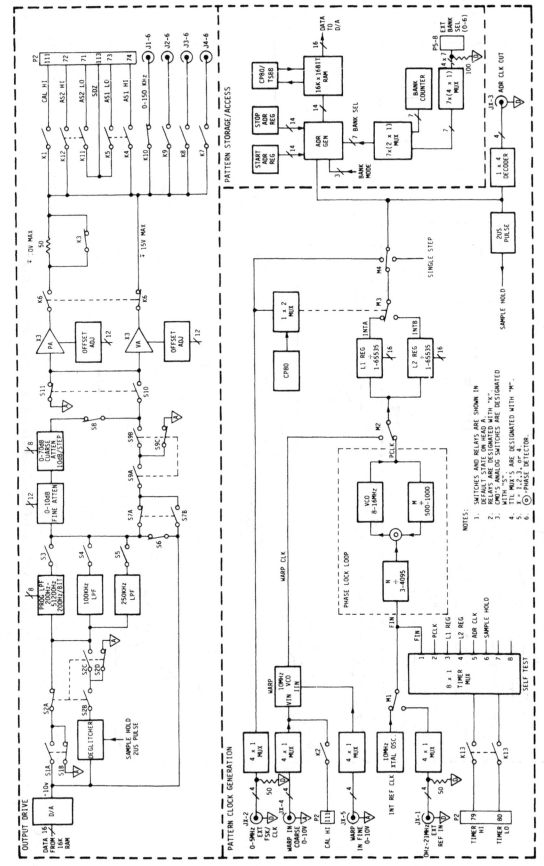

Figure 1.7: WS800 waveform synthesizer functional block diagram

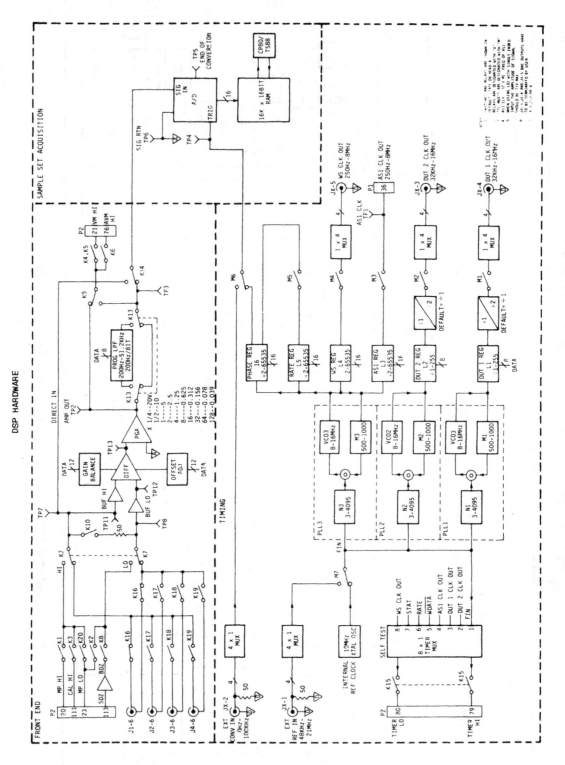

Figure 1.8: WD 800 waveform digitizer functional block diagram

(VCO), VCO 3, could be set to 257 times 50 kHz, or 12850 kHz, and register L5 set to divide by 257. The result, 50 kHz, is delayed a programmable amount by L6 and used as the A/D strobe.

An external sine wave generator could then be locked by setting divider L4 to 256, producing an output of 50 kHz * 257/256, or 50.1953125 ks/s.

By using one PLL as a common clocking source for the dividers, this scheme minimizes any jitter that may exist in the PLL itself. The PLL in this example serves primarily as a programmable crystal clock at 12850 kHz. If the phase comparator (the little double circle) were optimized for 16 to 20 kHz input signals, a VCO output of 12850 kHz would be produced by setting divider N3 to 600 and feedback divider M3 to 771. When the comparison inputs are identical in frequency and equal in phase, this particular comparator design generates a steady DC control voltage that tells the VCO to continue doing exactly what it is already doing. If the VCO begins to drift a little, the comparator will increment or decrement the DC control voltage to correct the drift.

This simple example introduces two concepts that will be explored later. The first is that of the *unit test period* (UTP). This is the *common* or *joint* period for all signals and sampling. Here, the UTP is 257 signal cycles, or 256 samples. The second concept is that of using relatively prime rate ratios (those with no common factors) to produce uniform, high-resolution sampling without resorting to high sampling rates or incremental time-delay circuits. In this simple example, what we have just done is to "walk" the samples forward through a repetitive waveform, *so that at the end of 1 UTP the vector appears to sample one signal cycle at 256 equally spaced points.*

DSP-Based Test Advantages Summarized

The DSP approach introduced provides a number of benefits in comparison with traditional analog test approaches. In manufacturing, increased test throughput is one of the most important. Reduced switching and settling time is one of the reasons; another is that the device response is memorized and can be analyzed for many parameters without recalling the DUT. In addition, software instruments need not operate in real time. Computation can proceed while the device is undergoing a different test.

Coherence is another technique that provides higher throughput. It allows the programmer to collect and process only the minimum-size vector needed to provide the required accuracy. In later chapters, we will see how it also permits the use of multitone testing, in which a number of different tests can be conducted simultaneously. For complex test plans, such techniques can often provide a hundredfold increase in throughput compared to hardware-based analog test systems.

Question: What are the other advantages over analog hardware?

Answer: For AC and dynamic testing, DSP testing offers several benefits:

1. It is nearly always more accurate.
2. It is more repeatable, machine to machine.
3. Calibration is much simpler.
4. Maintenance is reduced.
5. DSP generally provides additional information along with the desired parameter. (A DSP peak detector, for example, tells you not only the peak value but also where the peak is located.)
6. It makes hitherto difficult measurements practical in volume production (e.g., phase and spectral distribution).
7. It can model the device, both ideally and with flaws, and thus show how the device should perform. This is a valuable aid in creating and verifying any test program.
8. It can assist the manufacturer in diagnosing device failures and help to spot trends.
9. It is extremely flexible. The test conditions or "fixtures" can be changed to something entirely different by just a few keystrokes.
10. It results in a general-purpose tester that is smaller, cheaper, and less power hungry than one built with conventional hardware.

Price of Using DSP

As a closing comment, I would like to note that the very flexibility of DSP systems, an asset to the skilled engineer, may well be a liability to the unskilled. Conventional instruments tend to be very forgiving because they are designed to do specific jobs. An unskilled operator can obtain useful answers without understanding the theory of the instrument or the mathematical nature of the measurements. Most engineers learn rather quickly how to operate new hardware instruments by reading the labels, poking a few buttons, turning a few knobs, and observing the results.

Not so with a DSP-based tester. It is rather like a combination lock with many dozens of numbers: The probability of getting the lock to open by trial and error is a very close approximation to zero. You need to know the combination in advance.

In a tutorial I gave overseas several years ago, the real "cost" of using DSP instruments suddenly came clear to one of the engineers midway through the day. He knew only a few words of English, but managed to express his revelation in a way that is all the more expressive for its simplicity. Leaping to his feet with a mixture of excitement and fear, he said, "But...this means...we must know something!"

Indeed we must. A DSP machine will do whatever we ask, nothing more. It is as clever, or as stupid, as we are. If we do not know how to test a device, the DSP machine will not know, either. Like a mirror, it reflects our own technical strengths and weaknesses.

By "something," this engineer meant knowledge far beyond just knowing how to operate the equipment and what was learned in the university. We must also know the physical and mathematical principles underlying each test; the design, behavior, and end use of the DUT; and the error sources inherent in the tests and those of the DUT. We must know what the tests are intended to show, and how the customer will use and interpret this information. In short, the real price of DSP testing may well be that it forces us to master the craft of engineering.

Chapter 2
Accuracy and Speed of Emulated Instruments

Chapter 2: Accuracy and Speed of Emulated Instruments

Conventional analog-based instruments are far less accurate in measuring dynamic (AC) parameters than in measuring DC. The Hewlett-Packard 3456A, 6-1/2 digit .transfer standard, for example, can measure DC voltage to an absolute accuracy of 15 ppm ± 2 counts, 90 days out of calibration. AC (RMS) accuracy, by contrast, is only 700 ppm ± 730 counts. The AC uncertainty is even greater for waveforms with high peak-to-RMS ratios.

The reason for this is important to our story, for it is not something inherent in the nature of AC parameters, but rather a weakness of the particular structure and circuitry used in conventional analog instruments, a weakness that digital signal processing (DSP)-based architecture avoids.

Traditional analog automatic test equipment (ATE) design focuses on a central A/D converter (ADC). By itself, this ADC is quite accurate and generally serves as the master DC voltmeter. To measure AC parameters, it is preceded by a detector circuit (Figure 2.1).

Ideally, the detector produces a steady DC voltage directly proportional to the specific waveform property under investigation. If the peak voltage of some waveform is 1.7 volts, for example, a peak detector should produce a steady output of (1.7C) volts, where C is a constant of calibration.

The detector originated as a means of producing a steady current to drive a galvanometer or a d'Arsonval meter movement. Today, the concept is still useful for dedicated meters with digital displays, because it allows the use of low cost integrating ADCs with inherently good long-term linearity and line frequency rejection. Analog ATE in the 1970s retained this concept because it was deeply ingrained, and fit with the prevailing view of test systems as assemblies of bus-controlled but otherwise conventional instruments.

However, it is the detector that is responsible for diminished AC accuracy. The detector also imposes a rather low limit on test speed by virtue of the long-time constant filter it employs to produce a smooth DC output. Speed is a separate issue, though, and we will consider the problem of accuracy first.

The reason that analog measurement is less accurate for AC than for DC is that nearly every AC detector employs at least one nonlinear function (e.g., square law, square root, logarithm, exponential, rectification, clipping, or multiplication by, or division by, another waveform). Poor accuracy is a result of the fact that analog circuits cannot reproduce nonlinear characteristics nearly as well as linear ones. While a good ADC can easily provide 0.005 percent (50 ppm)

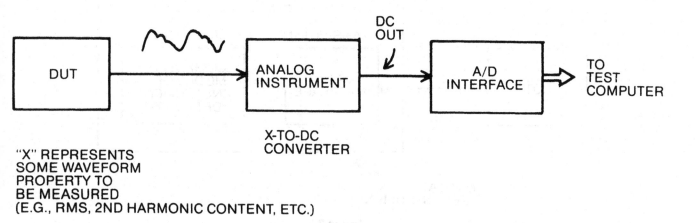

"X" REPRESENTS
SOME WAVEFORM
PROPERTY TO
BE MEASURED
(E.G., RMS, 2ND HARMONIC CONTENT, ETC.)

Figure 2.1

linearity, it is hard to find analog ICs or hybrid circuits that adhere to nonlinear characteristics more closely than about 0.1 to 1 percent (1000 to 10,000 ppm) over common ranges of frequency, temperature, crest factor, time, etc. Calibration of nonlinear analog functions in a production environment is also a problem.

Hardware Emulation

The first order of business in designing a DSP-based test system is to place the ADC ahead of the detection function, so that it receives the device under test (DUT) signal directly. There, it converts the waveform into a numerical vector and sends it on to the detector (Figure 2.2). This new scheme allows the detector to be modeled in software, using floating-point (FP) mathematics. There are no nonlinear analog functions preceding the ADC, only linear ones such as amplifiers, attenuators, or filters. Analog circuits can perform these linear functions with a high degree of precision.

By removing the detector from the analog domain, emulation thus makes it possible to provide AC and DC measurements of comparable accuracy. If each digitizer sample is absolutely accurate to (say) 50 ppm, then it is possible (in a coherent tester) to compute dynamic parameters to the same accuracy (i.e., to 0.005 percent, or 0.00043 dB). In fact, if the sample errors are not strongly correlated with the signal, statistical techniques can provide overall dynamic accuracy *superior* to that of the individual samples.

Integration versus Filtering for AC Measurements

Slow operation is a second disadvantage of conventional analog detection. This has nothing to do with nonlinear trans-

fer but is a consequence of the common practice of using a low pass filter to produce a steady DC detector output.

To perform true root-mean-square (RMS)-to-DC conversion explicitly, for example, an analog detector first squares the input waveform, then filters the result to produce the mean, and finally extracts the square root. The explicit form of RMS detection is shown in Figure 2.3.

The reader is probably aware that there are alternate and more favorable circuit configurations for RMS, based on an implicit algebraic solution. This is not relevant to the issue of speed, because the mean is generally produced in either form by RC low pass filtering. To minimize signal ripple, the filter must provide considerable attenuation at the test frequency. For this, the time constant must be much greater than one signal period, and the settling time, in turn, must be much greater than the time constant. Typical detectors take about 25 to 100 periods at the lowest-rated frequency to settle to 0.1 percent. Allowing a detector to settle longer than its rated time does not remove all error, since the steady state detector output always contains a certain amount of signal ripple.

The advantage of low pass filtering is that no synchronization is needed between the system functions. In hand held instruments, there is no significant disadvantage because the detector settling time seldom exceeds human response time, and the ripple error is usually less than the display error.

If we are willing to take the time and expense to synchronize the analog detector with the signal source, however, detection speed can be greatly increased, and ripple elimi-

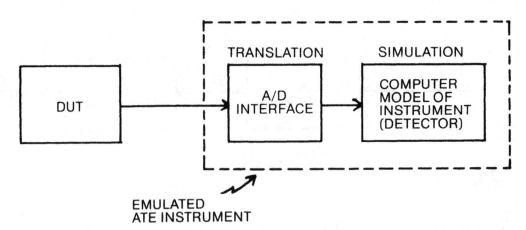

EMULATED
ATE INSTRUMENT

Figure 2.2

nated, by replacing the filter with a timed integrator. The interval of integration is adjusted to span exactly a whole number of cycles. Equations (2.1), (2.2), and (2.3) show the coherent form of three familiar measurements: DC, (absolute) average AC, and true RMS.

$$V(dc) = \frac{I}{P} \int_P V_{in} \, dt \qquad (2.1)$$

$$V\left(\begin{array}{c} abs. \\ avg. \end{array}\right) = \frac{I}{P} \int_P |V_{in}| \, dt \qquad (2.2)$$

$$V(rms) = \sqrt{\frac{I}{P} \int_P V_{in}^2 \, dt} \qquad (2.3)$$

In each equation, the letter P represents the so-called *test period*, or integration interval. For accuracy, P must be one or more whole periods of the DUT response signal. At the end of this interval, the integrator holds a steady voltage proportional to the magnitude of the detected waveform parameter. Filter settling is eliminated, and replaced by a relatively short—and programmable—"window" of integration.

Coherent Measurement

If the first order of business in DSP design is to replace all the nonlinear analog functions with software models, then the second order of business is to replace filtering with *timed integration* (i.e., using the correct equations). In DSP-based

testing, of course, integration cannot take the classical form, because emulated instruments operate on vectors and not on continuous waveforms. DC, absolute average, and RMS are computed by the discrete equivalents, as shown in Equations (2.4), (2.5), and (2.6).

$$V(dc) = \frac{I}{N} \sum_{I=1}^{N} V(I) \qquad (2.4)$$

$$V\left(\begin{array}{c} abs. \\ avg. \end{array}\right) = \frac{I}{N} \sum_{I=1}^{N} |V(I)| \qquad (2.5)$$

$$V(rms) = \sqrt{\frac{I}{N} \sum_{I=1}^{N} V(I)^2} \qquad (2.6)$$

The principle of coherence can be applied to analog circuits as well as DSP systems. Although the concern lies primarily with its application in DSP-based systems, the concept is generally easier to visualize in the analog domain. Coherent RMS measurement, for example, can be demonstrated by the analog equivalent of Figure 2.4.

Unit Test Period

DSP system coherence is more complicated than the above analog example suggests. For one thing, the signal is analyzed in discrete (vector) form, meaning that P must contain not only a whole number of signal cycles but also a whole number of sampling intervals. Sometimes, these numbers are

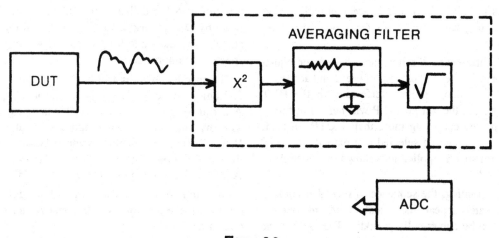

Figure 2.3

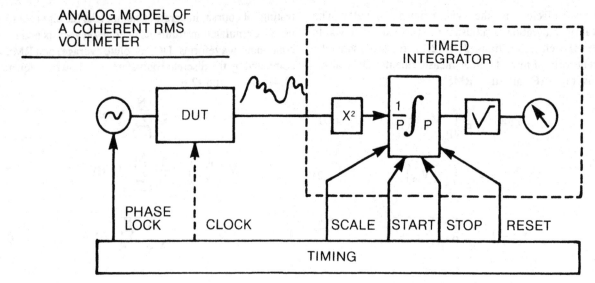

ANALOG MODEL OF
A COHERENT RMS
VOLTMETER

TIMED
INTEGRATOR

DUT

X² $\frac{1}{P}\int P$ √

PHASE
LOCK   CLOCK

SCALE START STOP RESET

TIMING

Figure 2.4

restricted by the algorithms to certain integers and, in all cases, are limited to special ratios. Finally, if the DUT is a sampling device (e.g., a switched-capacitor filter), its clock rate must also be coordinated with the others according to certain rules.

In the majority of commercial DSP applications, restrictions imposed by the DUT and/or the digitizer make it impossible to gather the required information from the DUT response in just 1 signal cycle. A key time interval in such cases is the unit test period (UTP), the interval over which unique information is spread. The UTP contains a whole number of signal cycles (M) and a whole number of sampling intervals (N), where M and N are relatively prime integers, that is, they have no common factor other than one.

By choosing different M/N ratios, the programmer adjusts the UTP to obtain the desired balance of speed and accuracy. For example, if random noise interferes with the DUT response, extending the test period P will help. But this is not done simply by collecting more than 1 UTP. Instead, a new M/N ratio should be selected that extends the UTP itself. Except for special applications shown in a later chapter, P = 1 UTP.

In coherent sampling, the sequence of recorded numerical values ultimately repeats itself, and an infinite coherent vector is seen to be numerically periodic. The UTP is the vector period, and corresponds to what is often called the *primitive* period in function theory.

To see why we must deal with the primitive or unit period instead of the signal period, consider the problem of testing a CODEC encoder (the serial ADC used in digital telephony). In standard telephone systems, the sampling rate is 8000 s/s, and the working audio (voice frequency) range is about 300 to 3400 Hz.

Assume that gain and signal-to-distortion ratio are to be measured with a "tone," or sinusoidal waveform, at a test frequency near 1000 Hz. The serial digital output is converted to parallel and collected in the "receive RAM" until a vector of N words has been collected. The vector is analyzed by DSP algorithms for the power of the fundamental signal component and for the total power of the remaining spectral components.

The encoder contains an 8-bit FP (*mu-law*) converter with 255 unique code levels. To get a reasonable picture of the A/D transfer curve, it is obvious that there should be at least as many samples. If the test signal is a sinusoidal wave, or "tone," there must be more samples because the amplitude distribution does not match the mu-law step distribution. AT&T standards require at least N = 400.

For our first experiment, let us try the following conditions: F_t is the test, or tone, frequency, and F_s is the sampling rate.

Trial 1:

$$F_t = 1000 \text{ Hz}$$

$$F_s = 8000 \text{ s/s}$$

$$P = 50 \text{ ms}$$

$$M = 50 \text{ cycles}$$

$$N = 400 \text{ samples}$$

At first glance, this may seem to do the job. But if you sketch the way that samples fall on the test waveform, you will see that, after eight samples, the pattern repeats. Subsequent groups of eight fall at the same relative points. At best, only 8 out of 255 steps will be evaluated, and we will not learn much about the ADC transfer characteristic. Worse, if the first sample falls at 22.5 degrees, only four unique locations will be sampled. For these reasons, 1000 Hz is not permitted for CODEC gain and distortion testing.

The problem above is that M and N are not relatively prime. When their ratio is reduced by the common factor, 50, correct values are obtained:

Trial 2:

$$M = 1$$

$$N = 8$$

$$UTP = 1 \text{ ms}$$

This tells us right away that the DUT information available with 1000 Hz is, at best, that obtained in just eight samples. The UTP is 1 ms. If P = 50 ms, this sparse information is simply repeated 50 times.

How can 400 samples be made to fall uniformly over the entire transfer curve? One way would be to set F_s, the sampling rate, to 400 ks/s. This is not possible with the CODEC, of course, which is designed to run at 8000 s/s. Even in devices that allow such a clock range, we would not want to shift F_s this dramatically because the dynamic parameters would be greatly affected.

Prior to the ready availability of coherent systems, a frequent solution to such problems was to offset F_t slightly, so that the input signal would drift with respect to the samples and cause each signal cycle to be sampled at different points. In the absence of synchronization, the uncertainty of sample location makes this scheme somewhat analogous (although not identical) to random sampling.

Given enough time, nonsynchronous, frequency-offset sampling provides very detailed information about the test device. With practical vector lengths, however, the uncertainty of sample location reduces measurement precision and reduces the repeatability from run-to-run or from machine-to-machine. Past experiments with CODEC gain and distortion suggest that it takes roughly 2000 random samples to give the same level of precision and repeatability that can be obtained from 400 correctly distributed samples. Another drawback is that sequence-dependent parameters (e.g., monotonicity and hysteresis) cannot be calculated from random samples.

The ideal sampling method is one that allows the programmer to select a sufficient, but minimal N, then distribute the samples over a controlled interval in a way that is informationally equivalent to uniform and sequential distribution over one signal period.

Coherent testing, with relatively prime M and N, is a systematic method of achieving this. It is informationally the most efficient method for unknown or arbitrary waveshapes, meaning that it maximizes the average information per sample for any arbitrary vector length. Like the previous method, it uses an offset F_t that makes samples fall at different points in subsequent cycles. With coherence, however, the locations are controlled and repeatable. Most important, if the individual signal periods are graphically superimposed, the resultant pattern will show N samples equiprobably distributed in time over one waveform cycle. This single composite cycle is termed the *primitive cycle*. Primitive spacing is 360/N degrees.

Most of the relations in coherence can be derived with three basic equations:

$$F_t = M * \Delta \qquad (2.7)$$

$$F_s = N * \Delta \qquad (2.8)$$

$$\Delta = 1/UTP \qquad (2.9)$$

where M and N have no common divisor other than 1. Δ is the so-called primitive frequency.

Dividing Equation (2.7) by Equation (2.8) gives the equation that describes the fundamental requirement for coherence:

$$\frac{F_t}{F_s} = \frac{M}{N} \qquad 2.10$$

These equations tell us how to solve the original problem. Since F_s and N are given, Equation (2.7) fixes the primitive frequency at 20 Hz. In coherent testing, all periodic components, whether desired or not, should be multiples of Δ. Therefore, F_t must be chosen from the series, ... 980 Hz,

1000 Hz, 1020 Hz, 1040 Hz, ... F_t must also meet the condition of Equation (2.10), however, leaving 980 Hz and 1020 Hz as the closest available frequencies to 1 kHz. AT&T standards further fix the range from 1004 to 1020 Hz, leaving only one valid answer. The numbers are as follows:

Trial 3:
$$F_t = 1020 \text{ Hz}$$
$$F_s = 8000 \text{ s/s}$$
$$M = 51$$
$$N = 400$$
$$UTP = 50 \text{ ms}$$
$$\Delta = 20 \text{ Hz}$$

Coherent Filtering

Suppose we want to know the amplitude of the fundamental component of a distorted 1 kHz signal. Given the rudimentary DSP measurement system shown so far, the desired measurement could be made by placing a 1 kHz bandpass filter ahead of the digitizer and computing the RMS value of the resulting vector (Figure 2.5).

Unlike the analog detectors discussed earlier, the filter is a linear function, and accuracy is not the issue. The problem this time is low speed. Because the filter in Figure 2.5 is required to produce a real-time waveform, it has real settling time, no matter how it is implemented. The settling time is substantially longer than one signal period. For simple ATE filters, a rough rule of thumb is 5 to 10 times the reciprocal of the three dB bandwidth, to settle to 0.1 percent. Flat-topped and "hard" wall filters take much longer.

However, the problem does not call for the filtered waveform. We are asked only to find its amplitude. Fortunately, frequency-selective amplitude measurement of coherent signals can be described by timed integrals, making it theoretically possible to perform the above measurement within 1 UTP.

Composite filter/measurement can be applied to the above problem in two ways. One is to replace the time-domain filter with a coherent hardware circuit whose output after one UTP is a DC voltage proportional to the second harmonic amplitude. The other is to digitize the unfiltered DUT output first and then use the software equivalent.

Coherent filtering of this kind is based on the fundamental operation of *waveform correlation*. There are a number

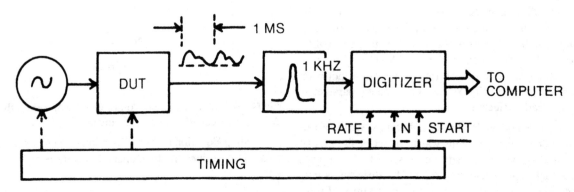

Figure 2.5 : Ineffective use of DSP

of excellent texts that provide a formal treatment of the topic, including those listed in the references. Rather than attempt to repeat this material here, I would like to take a different tack by developing the correlator and its filter variants from a visual, intuitive standpoint. This is particularly helpful in "testing," for a good DSP machine deals with mathematical functions as modular building blocks. It helps to be able to think of them as circuits rather than as equations.

Correlation

In its coherent form, correlation is defined by Equation (2.11)

$$R(\tau) = G \cdot \int_P A(t) \cdot B(t-\tau)\, dt \qquad (2.11)$$

When dealing with vectors, A and B may be functions of any variable the programmer chooses. To visualize correlation in circuit form, however, we may treat A and B as functions of time (i.e., waveforms). In Figure 2.6., variable τ is a programmable delay. G is the "gain," or scale factor.

In this hypothetical analog circuit, there is a single delay line, placed either in the upper or the lower input path. If the delay term τ is positive, the delay line is placed in path B so as to make B *lag* A. If τ is negative, the delay line is placed in path A so as to make B lead A (i.e., experience a negative delay). The magnitude of time delay is given by the magnitude of τ. Bipolar delay permits the extension of this correlation circuit to signals that are not periodic.

DSP systems often provide correlation as a directly available function, as well as using it internally to build other DSP routines. Two examples of syntax from the LTX DSP operating system are shown in Equations (2.12) and (2.13).

$$\text{MAT R = B CORRELATED WITH A,} \qquad (2.12)$$

and

$$\text{MAT R = B CORRELATED WITH A SHIFTED} \qquad (2.13)$$
$$-3 \text{ TO } +17$$

The prefix MAT is the BASIC notation that tells the compiler that the operands are not scalar variables, but matrices of numbers (in this case, vectors). The default delay is zero,

so in the first operation vectors A and B are compared without shifting. The answer is a single number (i.e., a scalar result).

Adding the *shifted suffix* causes the correlation computation to be repeated for each shift in the given index range. This time, the answer is a *vector* (in the above case, containing 21 correlation products).

For any given shift, correlation shows the degree to which two waveforms (or functions) are alike in shape. In *normalized correlation*, R is limited to a range of -1 to $+1$, regardless of the actual amplitudes of the two inputs waveforms. A correlation result of $+1$ indicates that the waves are identical in form. If $R = -1$, the waves have identical, but inverted, form. If $R = 0$, the waves are statistically unrelated for the specified shift.

To normalize correlation, the following scale factor is used:

$$G = \frac{1}{\text{RMS(A) * RMS(B) * UTP}} \qquad (2.14)$$

The term, *cross-correlation*, indicates that two different signals are compared. *Autocorrelation* is the process whereby the same waveform is applied to both inputs.

With zero shift, autocorrelation is always $+1$. As the delay is varied over its full range, autocorrelation varies in a way that is characteristic to the input waveform. If A is periodic, then R is also a periodic function (i.e., another waveform).

Exercise 1: Is the autocorrelation waveshape a function of time?

Exercise 2: What is the autocorrelation function of a sine wave?

Exercise 3: Sketch a wave whose autocorrelation is never negative.

Autocorrelation can be used as a form of low pass filtering. With a noisy sinusoid, the result is a fairly pure cosine wave, for example. But filtering of this kind takes considerable computer time and does not provide the kind of frequency selectivity best suited to DSP testing. Furthermore, measurement would have to be performed separately from filtering.

By combining two correlators with certain other functions, however, we can construct a very rapid and versatile filter/measurement unit that for tutorial purposes is called the *Fourier voltmeter* (FVM). To see how this works, we need to understand two simple but important principles that govern the correlation of sinusoidal waves.

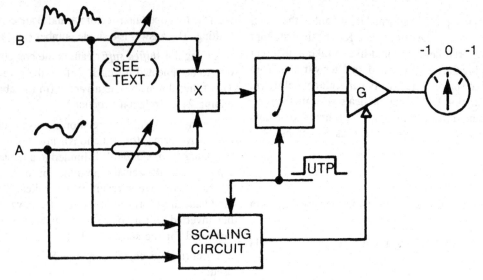

Figure 2.6: Analog model of correlation function

The first principle has to do with steady state sinusoids of different frequency. Assume that the two signals are applied to the previous correlator module of Figure 2.6, and let the letter P denote the interval of integration, or test period. In this experiment, P is not necessarily the UTP, but any interval of interest (Figure 2.7).

If P is infinite, the correlation between the two sinusoids is zero. For finite P, Fourier's first principle may be summarized as follows: (1) If A and B are sinusoids of unequal frequency, and (2) if interval P contains a whole number of *A* cycles and a whole number of *B* cycles, then (3) the cross-correlation is zero, *regardless of phase or amplitude*.

This is one of two ways to force zero cross-correlation between two sinusoids. *The importance of this principle to filtering is that it applies independently to each and every component of a complex wave formed by the linear addition of coherent sinusoidal components.*

To illustrate, start with the autocorrelation case in which sinusoid A is applied to both inputs, and R = 1. Now, linearly add a sinusoid of another coherent frequency to the lower channel without changing G. The correlation R is unaffected by this new component. The two A components correlate completely, while the cross-correlation of A with

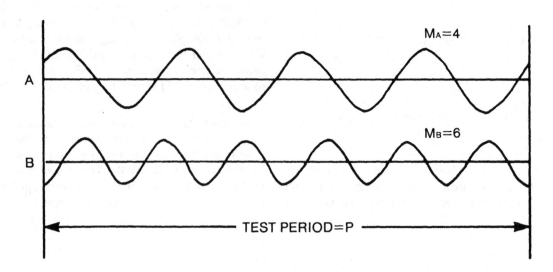

Figure 2.7: First Fourier principle

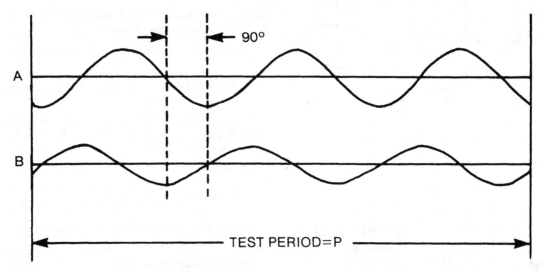

Figure 2.8: Second Fourier principle

the new component is zero. The total correlation is the combined value: +1.

The principle extends to any number of additive components where coherence holds. The practical result is that we can use a correlator in a coherent system to serve as an infinitely selective "bandpass" filter. Fourier's first principle makes it possible to construct a circuit in hardware or software that will ignore all frequency components in a wave, *other than the component whose frequency matches that of the reference input.*

Question: Is P (Figure 2.7) equal in time to 1 UTP?

The second principle concerns sinusoids of the same frequency (Figure 2.8): (1) If two sinusoids are equal in frequency, and displaced 90 degrees in phase, and (2) if there is a whole number of signal cycles, J, in interval P, then (3) the cross-correlation is zero, *regardless of aamplitude or the starting point of interval P.*

If both G and the reference amplitude are held constant, then a single correlator will indicate the relative strength of an in-phase matching component. The second principle allows a pair of correlators to measure the relative amplitude of the matching component, *regardless of its phase.*

Any steady state sinusoid locked to the reference frequency can be mathematically resolved into two parts: one that is in phase with the reference signal and one that lags by 90 degrees. If the reference is defined as a cosine wave, a locked, matching component can thus be resolved into a cosine part and a sine part. When two correlators are driven by quadrature reference signals, one output will indicate the relative strength of the cosine part while the other will indicate the relative strength of the sine part.

Individually, these outputs vary with the phase of the test component. But the power sum of the two outputs is independent of phase and is proportional to the power of the matching component. Phase independence is assured by the trigonometric identity,

$$\sin^2 x + \cos^2 x = 1 \qquad (2.15)$$

The phase relative to the cosine channel is given by the identity,

$$\text{phase angle} = \arctan \frac{\sin x}{\cos x} \qquad (2.16)$$

Fourier Voltmeter

The preceding pair of quadrature correlators forms the heart of a modular filter function called the Fourier voltmeter (FVM). The FVM combines the functions of an infinitely sharp bandpass filter, an AC voltmeter, and a phase meter. With a single FVM, we can measure the magnitude and phase of any arbitrary spectral component in a periodic waveform. By using a rank of thousands of parallel FVM units, each tuned to a different multiple J, a detailed spectral analysis is produced in a fraction of a second.

Figure 2.9 shows two varieties of FVM: (a) has rectangular output (cosine and sine) and (b) polar output (magnitude and phase). The polar unit is an extension of the rectangular form, incorporating circuitry or software to perform the previous trigonometric operations, plus the square root to convert power into magnitude.

The rectangular form can be made from a pair of quadrature correlators in which the scale factor is a constant,

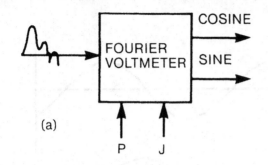

(a)

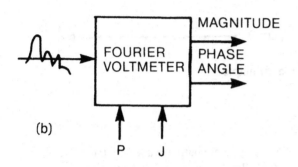

(b)

Figure 2.9

calibrated to denormalize R such that each of its two parts indicate the *actual amplitude in volts*.

The factor for normalization is given in Equation (2.17):

$$G \text{ (normalized)} = \frac{1}{RMS(A) * RMS(B) * UTP} \quad (2.17)$$

This produces a reading of +1.000 for perfect correlation, regardless of signal amplitude. If we wish R to indicate the *actual* component amplitude, E, we must multiply G by E:

$$G \text{ (FVM)} = \frac{E}{RMS(A) * RMS(B) * UTP} \quad (2.18)$$

In a DSP system, the FVM is implemented as a software routine, and the reference cosine (or sine) has unit amplitude. Amplitude means peak, of course, so its RMS value is $\sqrt{2}/2$. The RMS voltage of the signal component is $E\sqrt{2}/2$. By substitution

$$G \text{ (FVM)} = \frac{E}{\left(\frac{\sqrt{2}}{2}\right)\left(\frac{E\sqrt{2}}{2}\right) * UTP} \quad (2.19)$$

Therefore,

$$G \text{ (FVM)} = \frac{2}{UTP} \quad (2.20)$$

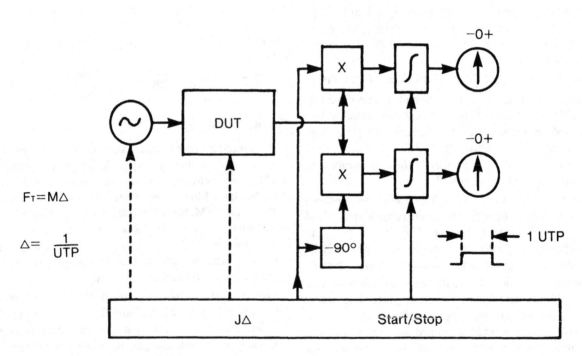

Figure 2.10: Analog equivalent of Fourier voltmeter

The rectangular form of the resulting Fourier voltmeter is shown in Figure 2.10. In coherent operation, none of the components in the DUT response will be detected except the matching component. This is the Jth multiple of the primitive frequency (i.e., $J\Delta$). To measure the fifth harmonic of DUT distortion, for example, M would be chosen according to the previous rules of coherence, and J would be set to 5M. Setting J to 0 gives the average or DC voltage.

Software Version of the FVM

For tutorial purposes, the FVM has been shown as an analog circuit; in an actual DSP system, of course, it is an algorithm that operates on a vector.

Let the signal vector be labeled X and its individual elements $X(I)$. Integer I is the index or position, 1, 2, 3, . . ., N. The vector equivalent of the analog FVM is performed by multiplying X by a cosine vector C and by a sine vector S. The individual sums of these two vector products are the cosine part and the sine part, respectively in Equations (2.21) and (2.22).

$$\text{cosine part} = \frac{2}{N} \sum_{I=1}^{N} X(I) \cdot C(I) \qquad (2.21)$$

$$\text{sine part} = \frac{2}{N} \sum_{I=1}^{N} X(I) \cdot S(I) \qquad (2.22)$$

In DSP computations, time is defined by index position, the analog scale factor 2/UTP is replaced accordingly by its *index* equivalent 2/N.

Vectors C and S contain exactly J cycles. They may be computed as part of the FVM algorithm, but it is faster to precompute them and store them in memory. In some machines, a reference vector pair is computed the first time a new value of J is encountered in a program, then stored for subsequent use as the program is repeated.

While the polar form is obviously convenient for phase or absolute voltage, the rectangular form of the FVM is faster and more accurate because there is less computation. The test engineer is encouraged to take advantage of this by working with signal *power* whenever possible, rather than amplitude (voltage). Where performance is expressed in dB ratio, for example, use the power form (Equation (2.23)) instead of the voltage form (Equation (2.24)).

$$\text{Number of dB} = 10 \log \left(\frac{P2}{P1} \right) \qquad (2.23)$$

$$\text{Number of dB} = 20 \log \left(\frac{V2}{V1} \right) \qquad (2.24)$$

If power in watts is required, remember that the FVM reports *amplitude* (i.e., peak voltage). Power computed from this is *twice the amount of the true power*. This cancels out in ratios, but do not forget to divide by 2 for absolute power calculation:

$$\text{Average sine wave power} = \frac{\text{peak power}}{2} \qquad (2.25)$$

Orthogonal Signals and Fourier Voltmeters

Coherent hardware was originally introduced so that integration could be used as a high-speed (and mathematically correct) replacement for analog filters and time-constant circuits. But coherence provides an additional benefit when two or more sinusoidal components appear in the response: It allows them to be statistically *orthogonal*.

For complex functions and waveforms, being orthogonal means having the same cross-correlation as two sinusoids at right angles, namely, zero. Two vectors are orthogonal if the sum of their index-by-index products is zero. Whether we deal with waveforms or vectors, though, the important thing is that orthogonality produces a unique relationship between the two quantities:

1. The two waveforms (or vectors) are statistically independent.
2. Each conveys separate, unique information.
3. If linearly added, they may be later separated without ambiguity.
4. If linearly added, the power of the resulting signal is the arithmetic sum of the individual powers.

The first two properties are important in communications theory. In testing, they permit the application of many different test signals at once and allow different DUT properties to be examined simultaneously.

Property 3 makes it possible to separate and measure each of the individual signal and distortion components accurately, even if there are thousands. Furthermore, since random noise is poorly correlated with coherent signals, true noise can be distinguished and measured separately from DUT distortion and quantization components.

Property 4 is deceptively simple, but vital. Total power *cannot* be computed by simple addition unless the components are orthogonal. It is this property that allows us to compute spectral line power directly from the rectangular FVM. If there is a separate FVM for every nonzero component in a given spectral zone, property 4 says that the total power is the *sum of the squares* (SSQ) of all the FVM outputs if, and only if, the measurement is coherent.

Fourier observed two centuries ago that sinusoidal orthogonality was completely ensured by just two conditions, repeated here for emphasis:

1. Sinusoids of different frequency are orthogonal over a given interval if each produces a *whole number* of cycles, regardless of the number of sinusoids or their amplitude or time relationships.

2. Two sinusoids of identical frequency are orthogonal over a given interval if they produce a *whole number* of cycles *and* the two waves are displaced by one-quarter cycle.

Coherence automatically establishes an interval (the UTP) that ensures the whole number conditions required in both cases.

DFT and FFT

The FVM was introduced as a tutorial concept to stress the modular function, not the computational procedure. You may see "FVM" in the actual list of DSP commands, or the algorithm by which it is implemented: the discrete Fourier transform (DFT). Two forms in current use are shown below. The prefix, MAT, an abbreviation of "matrix," is common to BASIC-like languages.

$$Y = \text{FVM} (X, J, N) \qquad (2.26a)$$

$$\text{MAT } Y = \text{DFT } J (X) \text{ sampled at } N \qquad (2.26b)$$

J is the multiple of Δ that you wish to measure. X is the label of the time series to be analyzed (e.g., the vector from the digitizer), and Y is a list of computed results. N is the num-

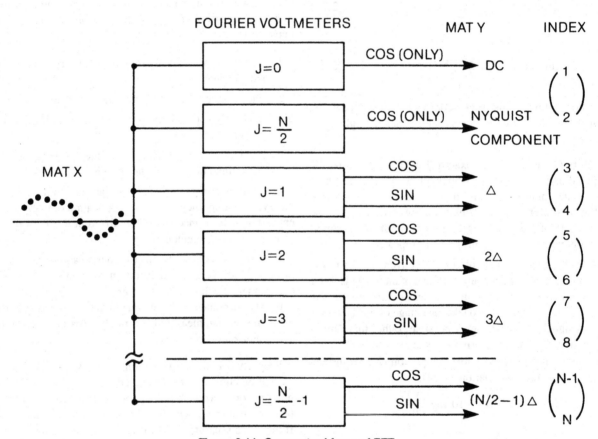

Figure 2.11: Conceptual form of FFT

ber of samples in the UTP. A single FVM or DFT operation can work correctly for any N.

Often we wish to know the amplitude and/or phase of many components. If N happens to be a power of 2, there is an algorithm that effectively applies a full set of FVMs at once: the *fast Fourier transform* (FFT).

$$Y = FFT (X) \qquad (2.27a)$$

or,'

$$MAT \ Y = FFT (X) \qquad (2.27b)$$

All M_i up to N/2 are evaluated, and the dimension of X is understood to be N.

The FFT is slower than a single DFT, but far faster than using a separate DFT for every spectral line. It provides N/2 FVMs, plus a DC voltmeter (Figure 2.11).

The FFT is thus the software equivalent of a spectrum analyzer that is sensitive to both magnitude and phase. If only magnitude is important, those answers are provided by the composite command,

$$MAT \ Y = MAG (FFT (X)) \qquad (2.28)$$

If you would like to know both phase and magnitude, follow the rectangular FFT with a polar conversion:

$$MAT \ Y = FFT (X)$$
$$MAT \ Z = POLAR (Y)$$

For simplicity, these operations may be nested in our tutorial system:

$$MAT \ Y = POLAR (FFT (X)) \qquad (2.29)$$

Synthesis

The FFT turns time information into frequency information. To create a vector for the waveform synthesizer, this process can be reversed. We start by defining the complex spectrum of the desired waveform, and then transform that into a time-domain vector by the *inverse* FFT. If F represents the frequency vector and T the time vector, the syntax might look like this:

$$T = INV \ FFT (F) \qquad (2.30a)$$

or

$$MAT \ T = INVERSE \ FFT (F) \qquad (2.30b)$$

This operation, like the forward version, works correctly only for selected values of N (most often, only powers of 2). This applies not only to the number of samples fed to the FFT but also to the number in the primitive period. If the UTP does not contain the right number, you cannot simply truncate the vector or fill out a short one by adding extra zeros or duplicate samples. These techniques are sometimes shown in texts, but they do not produce coherence or restore lost orthogonality. In such cases, we should alter F_s, F_t,

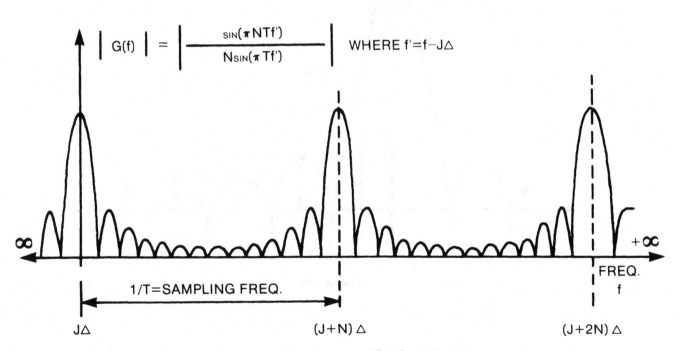

$$\left| \ G(f) \ \right| = \left| \frac{\sin(\pi NTf')}{N\sin(\pi Tf')} \right| \quad \text{WHERE } f'=f-J\triangle$$

1/T=SAMPLING FREQ.

$J\triangle$ $(J+N)\triangle$ $(J+2N)\triangle$

FREQ. f

Figure 2.12: Swept response of Fourier voltmeter

and/or M to permit the proper N. In the few cases where that is not possible, use the DFT.

Frequency Leakage

FVM measurements of either kind are precise only if all the components to be analyzed or synthesized are whole multiples of Δ. If the input stimulus contains only whole multiples, then all output harmonics, all intermodulation (IM) products, and all quantization distortion components are also whole multiples. But what if some input component is not a multiple?

Random noise is usually not a problem, because the correlation of noise with the reference frequency of the FVM is very small with large N. The situation that is most likely to cause trouble is the one in which the programmer has overlooked a component which is periodic, but which has no "bin" or FVM location itself. For example, suppose a telecom filter program is written for North America, using a primitive frequency of 20 Hz and a UTP of 50 ms. This provides an FFT bin to test the filter rejection at 60 Hz. The program is then sent to Europe, where the notch is tested at the power frequency of 50 Hz, midway between two bins. First, the measurement will be wrong. Second, an applied 50 Hz component will "leak" into every bin in the FFT spectrum and contaminate the other spectral measurements.

Leakage occurs because sinusoids that do not have a whole number of cycles over the integration interval are no longer orthogonal to the FVM reference frequencies. Usually, the amount of leakage is greatest in the bins closest to the offending tone. To see this, let us "sweep" an FVM input with a continuum of input frequencies and record the magnitude of the response. The response curve is shown in Figure 2.12 and is repeated with more detail in Figure 2.13.

All components that are whole multitudes of Δ fall into the nulls, except the one whose multiple is J (i.e., the one to be measured). If Δ were 20 Hz, for example, a tone at 50 Hz falls halfway between the nulls.

In the FFT, there are $N/2 + 1$ of these curves, one for each FVM, displaced horizontally by the primitive frequency. If Δ is, say, 20 Hz, then FVM 1 is tuned to 20 Hz, FVM 2 to 40 Hz, and so on. These are the bin frequencies. The bin numbers are integers from 0 to $N/2$. Bin 0 contains a single number, the DC voltage. Bin $N/2$ also contains a single number, the magnitude of the cosine component at the so-called Nyquist frequency. (The sine part is lost in sampling.)

Bins 1 through $(N/2 - 1)$ contain *pairs* of answers, either the cosine-sine parts, in the rectangular form, or magnitude-phase, in the polar form. Bin 1 holds the information about the primitive frequency, which should not be confused with the fundamental component of F_t. That, you recall, is the Mth multiple of Δ, and its information is found in bin M. Altogether, there are $N/2 + 1$ bins produced by the FFT and exactly N parts, or individual answers.

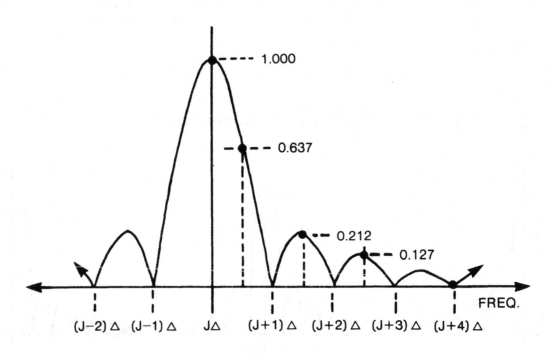

Figure 2.13

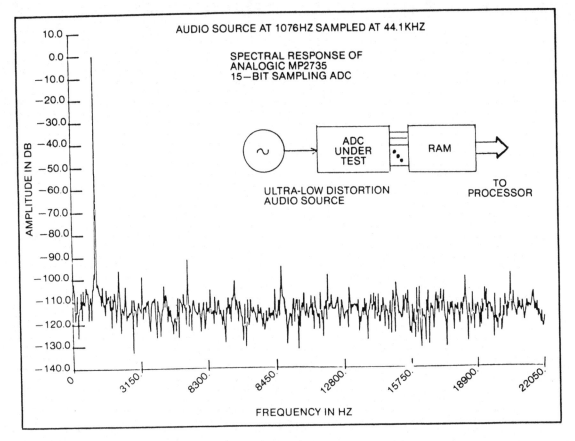

Figure 2.14

Suppose the primitive frequency is 20 Hz and there are no valid signal components at 40 or 60 Hz. Ideally, bins 2 and 3 would be empty. If a component at 50 Hz ($= 3\Delta/2$) is present, however, bins 2 and 3 will contain leakage components. There will appear to be signal components at 40 Hz, 60 Hz, and, to some extent, at all multiples of Δ. Figure 2.13 shows that the false component at 60 Hz has an amplitude 0.637 of the 50 Hz amplitude. In like fashion, the 40 Hz bin, bin 2, responds with the same false measurement. The next outer pair of bins reads 0.212 of the 50 Hz amplitude, the next outside these reads 0.127, and so on, following the function called Dirichlet's kernel. (This is not the same as the sync or sine x/x function.)

In many tests, the problem is that the other bins are not empty but contain components that you are trying to measure. The power of the leakage component and that of the valid component in each bin are no longer additive because they are not orthogonal, so there is no easy correction if leakage has been allowed to take place. A technique of shaped time windows ("windowing") is shown later, but this primarily helps strong lines. It is far better to choose F_s, M, and N such that all expected frequencies are multiples of Δ.

Graphical Example of FFT Application

Figure 2.14 shows the result of applying the MAG FFT to a vector or 1 UTP obtained from a commercial ADC (Analogic MP 2735). The input is a coherent sinusoid produced by driving a 16-bit DAC with a sinusoidal digital pattern at 1076+ Hz. The D/A output smoothed by a very low distortion bandpass filter and was verified to have no harmonic component above -115 dB. For this experiment, N = 1024, and the A/D sampling rate was set at 44100 s/s, a standard clock rate used in compact disc (PCM audio) recording.

Exercise: Determine the value of M and the exact value of F_t used in the experiment.

Chapter 3
Noncoherent Sampling

Chapter 3: Noncoherent Sampling

Digital signal processing (DSP) testing involves two kinds of sampling. The system digitizer is designed to sample coherently, but the device under test, if it is a CODEC, switched capacitor filter, mixer, or A/D or D/A converter, is designed to sample noncoherent waveforms (e.g., speech, music, or video signals). In Chapters 3 and 4, we will contrast these two forms of sampling and relate them to practical DSP testing.

Let us begin with the more general and more familiar case, namely, *noncoherent sampling*. As an example, consider a speech waveform that has been band-pass filtered for transmission over telephone lines. Its spectrum rolls off sharply below 300 Hz and above 3400 Hz, and the power is nil at DC and above 4000 Hz.

If this time-varying waveform is sampled at 8000 s/s (Figure 3.1c), no information is lost. The original spectrum is completely retained. Sampling merely replicates the spectrum indefinitely, as mirror *images* (Figure 3.1d). The spectrum becomes *periodic*, with a period of F_s.

Reconstruction

If a function is sampled only at discrete points, it would seem that the variations or ''wiggles'' between the points are lost. In a local sense, this is true; given only two adjacent samples, there is no way of knowing how the waveform actually varies between them. The full sample set, however, contains all the information needed to *reconstruct* the complete, continuous function.

Figure 3.1d suggests the principle by which the complete time function can be restored: Suppress all portions of the periodic frequency spectrum except the original portion. This forces the time function to be continuous and to follow the original shape. The filter that removes the unwanted spectral sections, and thereby connects the time samples, is called the *reconstruction filter*.

True reconstruction, meaning connecting the samples with a smooth, continuous curve, requires classical analog filtering. Digital reconstruction filters can approximate this curve by adding extra, interpolated points between the initial sample points. In real-time digital reconstruction or interpola-

tion, this means that the output clock rate is increased with respect to the primary sample rate. Filters of this kind are sometimes called *oversampling filters*.

Time and Spectral Vectors

Since a sampled time-varying signal is discrete along the horizontal axis, but continuous along the vertical axis, it can be directly and completely expressed as a *numerical vector*. Unfortunately, if the original signal were truly nonrepetitive, the vector would have to be infinite in length to capture all the information.

The DSP processor cannot deal with infinite vectors, of course. But we might inquire if, instead of an infinite time vector, the *spectrum* could be computed and stored. Even though infinite in length, the time vector is *discrete*, which means that its spectrum is therefore *periodic*. Only 1 period need be stored, and this has a finite frequency span. Does this provide a way to store an infinite signal in a finite space?

Sorry, but this does not solve the problem. Even though the time waveform is discrete, its spectrum is not (Figure 3.1d). Any nonperiodic time signal, discrete or continuous, has a *continuous spectrum*. The DSP processor cannot represent a finite-length, *continuous function* any better than an infinite-length discrete one.

The solution is to sample the spectrum as well as the time series. That will make it both discrete and finite and, thus, representable by vector. Sampling the spectrum, however, forces *periodicity* on the original time-varying waveform. In that case, we could just as well store one period of the time signal as one frequency period of the spectrum. In brief, there is an informational equivalency between the time domain and the frequency domain. You cannot restrict the information in one domain without restricting it in the other.

The result is that the mathematics of DSP testing forces us to employ test signals and spectra that are simultaneously periodic and discrete. Instead of fighting this, the modern DSP designer turns this to the fullest advantage by making the system *coherent*.

But this is getting ahead of the story. The DUT, even if it employs sampling, is usually not memory limited. A

EH0258-4/87/0000/0035$01.00 © 1987 IEEE

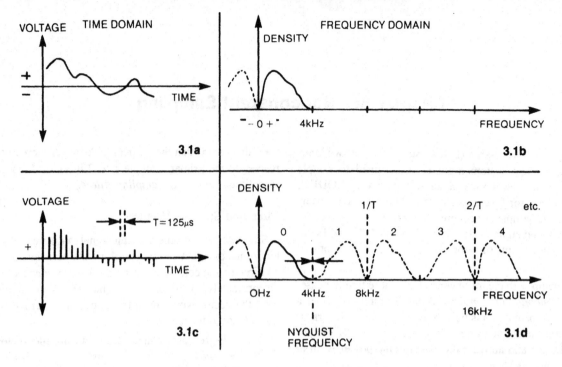

Figure 3.1: Spectrum of sampled signal

switched-cap filter, for example, does not care whether the signal repeats or not. DUT sampling, in contrast with that of the DSP-based tester, is designed to go on for years. It produces output signals whose frequency *spectra* are *periodic but continuous.*

Imaging and Noncoherent Undersampling

All sampled signals, periodic or not, have frequency spectra that are periodic over the sampling frequency, F_s. In sampling or reconstruction, however, only half of the period need be used; the remaining half is simply the reverse of the first half in sequence and phase. The half-period from DC to $F_s/2$ is often called the *Nyquist interval.* It is the default region of the DSP processor for most fast Fourier transform (FFT) and discrete Fourier transform (DFT) routines.

There is no requirement to sample in the Nyquist zone. There are many situations, in fact, where it is advantageous to use some other spectral region. A signal confined to the region from 8 to 12 kHz, for example, can still be sampled at 8000 s/s. The resultant spectrum would look the same as Figure 3.1d.

The DSP processor has no idea which signal is the "real" one, but simply reports the primitive values (i.e., the components in the Nyquist region).

This is a good illustration of *undersampling.* As shown in Figure 3.1d, any one of the 4 kHz-wide zones could contain the original analog signal.

Heterodyning and Reconstruction

We have the same freedom in reconstructing signals as in sampling them: An analog waveform can be synthesized in any of the half-period zones. You could sample a signal in the 16 to 20 kHz zone, for example, then "play back" the sample set through an 8 to 12 kHz bandpass fiter. All the original frequencies would be shifted downward by 8 kHz.

Question: What would happen if a band-limited signal in the 4 to 8 kHz zone were shifted in the above way to 0 to 4 kHz?

This technique is known as *heterodyning* in RF applications, where the local oscillator acts as the sampler. It is also used in frequency division multiplexing (FDM) as a means of assigning different frequency zones to signals that originally all have the same frequency range. An FDM signal is detected and demultiplexed by the same technique.

Please note that heterodyning or undersampling does not relax the requirements for low sampling jitter and aperture uncertainty or for low D/A glitch area. These requirements are determined by the actual test frequency, not the sampling frequency.

Note also that the signal to be sampled—if its spectrum is continuous over its band—must not spill over into an adjacent zone. If that happens, images will overlap and become ambiguous. Images that interfere are known as *aliases*.

Problems with aliasing are much easier to avoid in testing, where the signals have line spectra instead of continuous spectra. It is all right to apply frequencies in several different zones simultaneously as long as you do a little bookkeeping to keep track of the image locations and avoid aliasing. The relationships are given below.

Rules of Imaging

Let f^o represent any frequency, or its image, in the zone from 0 Hz to $F_s/2$. The frequency $F_s/2$ is the Nyquist frequency, and the zone 0 to $F_s/2$ is the Nyquist region. In coherent testing, the spectrum that appears in this zone is called the *primitive spectrum*.

Let f^1 represent the image of f^o that occurs in the zone, $F_s/2$ to F_s; let f^2 represent the next highest image, then f^3, and so on. These images (one of which is "real") are related as follows:

$$f^1 + f^o = F_s \qquad f^2 - f^o = F_s$$
$$f^3 + f^o = 2F_s \qquad f^4 - f^o = 2F_s \qquad (3.1)$$
$$\text{etc.} \qquad\qquad \text{etc.}$$

Sampling Rates and Spacing

You will often hear that signals should be sampled at "more than twice the highest frequency," but this results from a confusion between two similar-sounding relationships: Shannon's sampling theorem and Nyquist's sampling limit. The sampling theorem describes a *sufficient condition*, not a necessary one:

If a function of time f(t) contains no frequencies higher than W cycles per second (Hertz), it is *completely determined* by giving the value of the function at a series of points spaced 1/(2W) seconds apart. (See C.E. Shannon, "Communication in the Presence of Noise," *Proc. IRE*, Vol. 37, p. 10, 1949.)

Now there are several things noteworthy about this theorem:

1. It says "completely determined." This is the real thrust of the theorem: Even though sampling misses infinitely more points than it catches, we can nonetheless preserve *all* of the information about the signal. It assures us that completeness is possible.

2. It does not leave us guessing about how completeness is achieved. It gives a practical condition that is sufficient for any and all time-varying functions.

3. The theorem is significant in what it does *not* require: It makes no mention of the phase or time origin of the samples. This is a blessing not only in communications, but in DSP testing as well. If the DUT has unknown phase shift, that will not prevent us from measuring either the magnitude or the phase. In reconstructing the analog signal from its vector, the absolute phase of the reconstructed signal is not affected by the phase of the sample set relative to the signal.

4. It implicitly establishes the condition almost universally assumed in DSP texts: regular sample spacing over the window of computation: no gaps, no jitter, no FM or PM. It is *not* sufficient to use an irregular sample train whose *average* rate is 1/2W. Irregular sampling can capture all the information at sufficiently high rates, but the analysis is made more complex; you must know where the individual samples occur in time. For waves from stochastic sources, uniform sampling requires the least number of samples. Over the universe of all waveforms, uniform spacing provides the most information.

5. In the context of Shannon's article, the theorem pertains to the limiting case of classical functions, f(t), where t ranges from minus to plus infinity. For vectors of *finite* length, the theorem must be amended to read, "... a series of regularly points spaced more closely than 1/2W." This is true even if the vector spans a precise period of an infinitely periodic wavetrain. In the periodic case common to testing, just *one extra sample* will do the trick.

Nyquist's Limit

Shannon's theorem is often confused with Nyquist's limit, which, by contrast, expresses a *necessary* condition, pertaining to the *bandwidth* of the signal, *not* the highest frequency.

$$F_s > 2W \qquad (3.2)$$

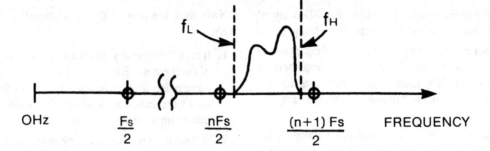

OHz $\dfrac{F_s}{2}$ $\dfrac{nF_s}{2}$ $\dfrac{(n+1)\,F_s}{2}$ FREQUENCY

Figure 3.2

If you go back to the previous examples of imaging, you will see that this rule is not violated. Every one of the signals or images is less than 4 kHz in width, thus satisfying the inequality.

Nyquist does not say how *much* higher F_s should be, but an exact expression is easily obtained by using the previous rules of imaging.

Universal Rule for Noncoherent Sampling

If the total energy of a signal is contained in a continuous spectrum of width $W = (f_H - f_L)$, an F_s must be chosen such that the interval f_L to f_H falls within two adjacent harmonics of $F_s/2$ (see Figure 3.2).

That is, if

$$f_L > \frac{nF_s}{2}, \quad \text{then} \quad \frac{(n+1)F_s}{2} > f_H \qquad (3.3)$$

By algebra, these two inequalities may be rearranged to give what we will call the *universal rule of sampling*:

$$\frac{2f_L}{n} > F_s > \frac{2f_H}{n+1} \qquad (3.4)$$

Where n is an integer ≥ 0. It is the image or zone number.

If $n = 0$, the condition is said to be the *low pass* case. Equation (3.4) is reduced to Shannon's condition of sufficiency. If $n = 1, 2, \ldots$, the condition is said to be the *bandpass* case.

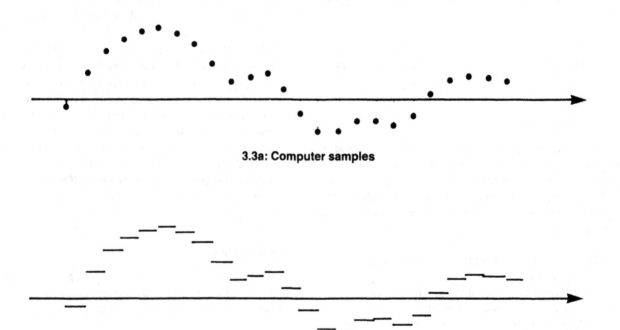

3.3a: Computer samples

3.3b: Samples at output of ideal D/A converter

Figure 3.3

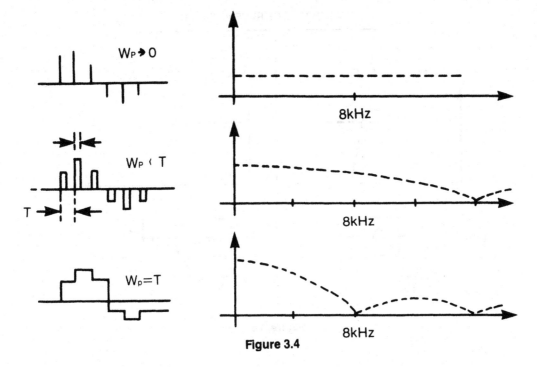

Figure 3.4

The bandpass case is not always possible, and in any event is limited to a few zones for F_s. The highest usable value for n, here denoted n*, is

$$n^* = \text{integer}\left[\frac{f_L}{W}\right] \qquad (3.5)$$

Exercise 1: An 80 kHz subcarrier is amplitude modulated by audio signals up to 4 kHz. The result is to be sampled. Calculate the lowest theoretical sampling frequency. Compare results with Equation (3.2).

Exercise 2: Determine the minimum sampling frequency if n=2.

Sine-x-over-x Distortion and Correction

So far, we have considered only ideal, zero-width samples (i.e., those that represent true points in time or frequency). This is a perfectly valid assumption for vectors that originate and remain within the computer. It is reasonably valid for vectors that are produced by DSP system digitizers and most high-speed ADCs. When vectors are removed from the numerical world, however, and turned into real waveforms by real DACs, the samples become finite in width, and the spectral picture must be modified. The distinction is illustrated in Figure 3.3a and b.

If a time-varying signal is represented by uniformly spaced pulses whose width W_p is very small when compared with the spacing T, then the spectral images produced by sampling will all be of equal strength. This is illustrated in the top sketch of Figure 3.4, where W_p is small, approaching zero. The dotted rule is essentially flat, indicating that the images (not shown in this simplified sketch) extend indefinitely to the right, with undiminished strength.

The dotted rule in Figure 3.4 shows the relative amplitude of spectral components with respect to frequency. As W_p becomes wider, more power is put into the low end of the spectrum and less into the high end. The spectrum rolls off as frequency increases, dropping to zero periodically at $1/W_p$, $2/W_p$, $3/W_p$, etc. Not only does this affect the relative amplitude of different images, but—more critical to DSP waveform synthesis—distorts each of the individual images in amplitude and phase.

The resulting envelope is said to follow the "sine-x-over-x" curve, where x is a function of W_p and frequency. At any frequency f the spectral component beneath the curve is attenuated and phase shifted by the amount

$$\left|\frac{\sin(\pi f W_p)}{(\pi f W_p)}\right| \underline{/-\pi f W_p} \text{ radians} \qquad (3.6)$$

In many texts you will find this expression written with T as the variable, not W_p. This is incorrect except in the limiting case shown in the lower sketch, where the pulses have a duty cycle of 100 percent, and W_p therefore equals T.

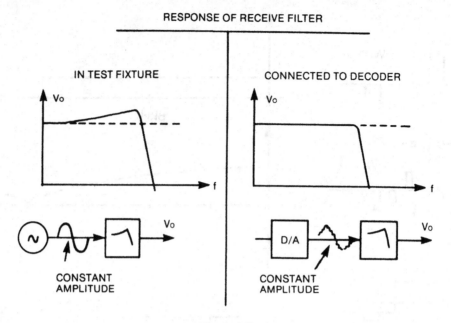

Figure 3.5

3-TONE WAVEFORM

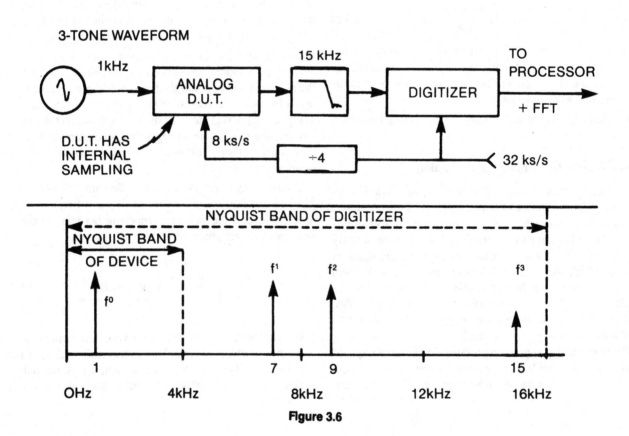

Figure 3.6

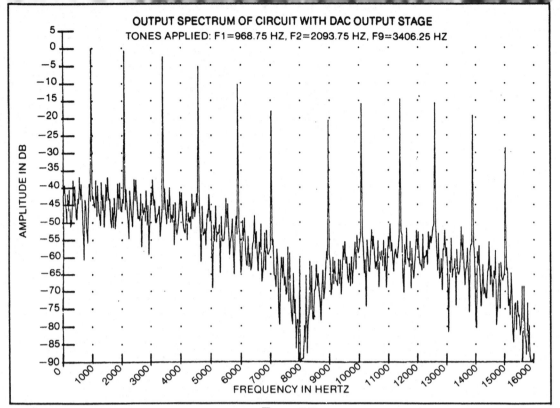

OUTPUT SPECTRUM OF CIRCUIT WITH DAC OUTPUT STAGE
TONES APPLIED: F1=968.75 HZ, F2=2093.75 HZ, F9=3406.25 HZ

Figure 3.7

Years ago this was a good assumption, but today's telecom, PCM audio, and RF DACs often use lesser duty cycles. Although reducing W_p reduces the signal-to-noise ratio, it allows the DAC to be de-glitched by shorting the output to ground during the transition interval. This is often called "return-to-zero" format (RZ). It also simplifies filter construction, which will be discussed later.

In all three pulsed waveforms, the tops are flat. The level represents the zero-order derivative—the function itself—at an instant of time defined by the *leading edge* of the pulse. This is known as *zero-order-hold sampling*. Informationally, all three forms are identical.

Receiver/Reconstruction Filtering

If sine-x-over-x rolloff is to be corrected by hardware reconstruction filters (the so-called "receive filters" in CODEC circuits), the filter curve must rise during the system passband so as to offset the sine-x-over-x droop, then drop sharply through the Nyquist frequency to remove the adjacent image (Figure 3.5). RZ format greatly simplifies the design of the filter.

Both imaging and sine-x-over-x rolloff are clearly seen in the following test of a real telecom circuit, an "analog" DUT

that turned out to be mostly digital. The DUT uses an internal ADC to sample the incoming signal at 8000 s/s, does some digital processing, then converts the digital stream back into analog form with a DAC. The DUT output is examined over the region from 0 to 16 kHz by a digitizer clocked at 32 ks/s (Figure 3.6a and b).

The digitizer anti-aliasing filter is set to cut off components above 15 kHz. Since the DUT itself samples at 1/4 the rate of the digitizer, an input f will show three significant images: f^1, f^2, and f^3. In this test, three frequencies of equal amplitude were applied simultaneously and coherently with the DUT clock. The nominal test frequencies are 1000, 2000, and 3500 Hz.

The spectrum is shown in Figure 3.7. On the basis of its information, see if you can answer the following questions:

1. Does the DUT contain a built-in reconstruction (receive) filter?

2. What is the duty cycle of the DAC inside the DUT?

3. What is the UTP?

4. What are the M/N ratios for the test frequencies?

41

Chapter 4
Coherence and Coherent Sampling

Chapter 4:   Coherence and Coherent Sampling

Conventional analog-based AC instruments are compromised in speed and accuracy with respect to DC voltmeters by two practical factors: (1) They generally employ nonlinear detector circuits and (2) they generally use low pass filters instead of timed integration. Digital signal processing (DSP)-based testing removes the first source of error by emulation (i.e., by executing the nonlinear part of the measurement function inside the computer using floating-point (FP) mathematics). DSP removes the second error source simply by obeying the correct (integral) equations.

The proper use of integral equations in dynamic measurements involves coordinating all generator and clock frequencies with the measurement window in the ratio of relatively prime whole numbers. In this tutorial, a test system programmed and controlled by such whole number ratios is said to be *coherent*.

The trivial case of coherence is the condition where everything runs at the same rate (i.e., where all ratios are 1:1). This is the equivalent of a mechanical system where all rotating elements are mounted on a common drive shaft. As the term is used here, however, coherence refers to the condition in which every element is allowed to run at a different rate, if required, yet be completely time coordinated in any way the programmer chooses. It is like a gear-driven mechanical system with any number of gears, no two of which need be the same size. The programmer is free to choose any number of gears, arrange them in any drive sequence, assign arbitrary sizes to the gears, and to rotate the individual gears prior to assembly so as to achieve the desired *phase* or timing relationship.

Coherence by PLL

The simplest nontrivial coherent test involves two frequencies, plus a measurement time window of 1 unit test period (UTP) duration. One of these frequencies is considered the primary frequency, F1, and the other is considered the secondary frequency, F2. F2 is generated from F1 by a *phase-locked loop* (PLL). The principle was outlined briefly in discussing the synthesizer and digitizer circuits in Chapter 1. In essence, the PLL is a circuit that provides the electrical equivalent of two gears (Figure 4.1).

What are the characteristics of such a system? In terms of frequency, the output shaft speed F2 is the primary (input) rate times the ratio of two relatively prime integers:

$$F2 = F1 * M/N \qquad (4.1)$$

Suppose F1 were the sampling rate of some analog-to-digital converter (ADC) under test, say, 20 Ms/s. We would like to test the ADC near the Nyquist frequency, such that N = 1024. A value of 511 is suitable for M, giving a test frequency F_t of 20 * 511/1024, or 9.98046875 MHz.

For simplicity, assume that the signal source is a sine wave oscillator that can be synchronized to a clock of the same rate. Let F2 represent the clock rate, so that it, too, will be 9.98+ MHz. The PLL enables the programmer to produce this exact frequency relative to a 20 MHz master clock by setting the feedback divider (forward multiplier) to M and the combined forward dividers to N = 1024.

Coherence by Parallel Division

Is there another way to produce these frequencies coherently? Yes. We saw the alternative in Chapter 1, where a common high frequency clock was used to drive two dividers set in the ratio N/M. Where practical, parallel division is superior to PLL synchronization by virtue of its simplicity, low jitter, and rapid response to new M/N settings.

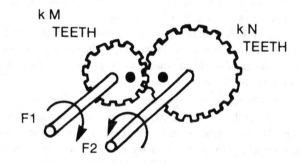

Figure 4.1

EH0258-4/87/0000/0045$01.00 © 1987 IEEE

In the ADC test problem just used, however, parallel dividers could be used only if the master clock were 511 * 20 MHz, or 10.22 GHz! Such a frequency is clearly impractical for driving dividers, especially those that contain large prime factors, like $511 = 7 \times 73$.

The PLL has the advantage that it can provide the *effect* of a 10.22 GHz clock without actually generating it. In Equation (4.1), this is numerically equal to the factor F1 * M. This is sometimes called the *implicit* clock rate:

$$\text{Implicit PLL clock rate} = F(\text{in}) * M \qquad (4.2)$$

A good coherent DSP system thus makes both approaches available. Parallel division is provided for those cases where the master clock rate is realizable. Good dividers provide highly regular sampling because the absolute jitter is carried down from the master clock, becoming relatively small when compared with the period of the divided clock rate. The PLL is needed whenever the master clock rate is so high that it must be implicity, not physically, generated. For low jitter and phase noise, the PLL will have a relatively long settling time.

Question: Can a *digital rate multiplier* be used in place of a PLL?

Vector Periodicity

In Figure 4.1, imagine that we place a mark on the rim of each gear, then set the gears in motion. With a stroboscopic light source, we observe the relative position of these marks at some instant of time t; let this be the state of the system at time t. As the system rotates, the gears will pass through the same state periodically, at $(t + P)$, $(t + 2P)$, $(t + 3P)$, and so on. P is formally termed the *primitive period*. It corresponds to the unit test period (UTP) in DSP testing.

If the gears rotate at a constant rate, then the UTP can be measured in units of time. The UTP can also be expresed in two ways that are invariant with speed, however. One is by counting the number of turns that one or the other of the two gears must make within the UTP. The other is to count the number of teeth that must pass a given point to return the gears to the starting state.

In DSP testing, the equivalent to turns counting is counting the number of signal cycles M, or the number of samples N. In a gear system where k=1 (i.e., where the individual numbers of teeth have no common divisor other than one), counting the number of teeth is equivalent to the number of *implicit* clock cycles within the UTP.

This number is normally of no interest by itself. However, dividing it by the time of the UTP gives the implicit clock rate. This is equal to the *effective sampling rate* in high-speed undersampling, the theoretical sample rate necessary to distribute N relatively by prime samples over 1 cycle of frequency F_t.

$$\text{Effective (implicit) sampling rate} = M(F_s) \qquad (4.3a)$$
$$= N(F_t) \qquad (4.3b)$$

Exercise 1: What is the effective sampling rate for the ADC example given at the beginning of this chapter?

Exercise 2: What is the effective rate in the CODEC problem where $F_s = 8000$ s/s and $F_t = 1020$ Hz?

Amount of Information in a Vector

The amount of information available from a sampled waveform is maximized when M and N are relatively prime. With relatively prime ratios, the amount of information is proportional to N and is *independent of M*.

Since there is no inherent upper limit to M, the above statement tells us that there is no lower limit to F_s in coherent testing. If a periodic waveform has no harmonics over the fiftieth, for example, any value of N over 100 will suffice to capture the total signal information (e.g., 128, if the FFT is to be used). These samples may be spread out over an arbitrarily long UTP, so as to provide an arbitrarily small F_s. The only requirement is that the product, $M(F_s)$ equal the product, $N(F_t)$, where M and N are relatively prime.

This is not possible with conventional, noncoherent sampling, where F_s cannot be less than the Nyquist limit, 2W. When the signal is completely periodic, though, it becomes possible to collect 1/Mth the information each cycle, until all M cycles have been sampled. A prime M/N ratio ensures that each cycle contributes unique, independent information.

With coherent sampling, it is not the actual sample spacing T that is important, but the *effective*, or *implicit*, or *primitive* spacing Tp. This is the spacing obtained by shuffling the vector samples so as to produce a single, *primitive*, cycle of the test waveform (i.e., one whose wavelength equals the vector length).

If Tp is expressed in degrees, the formula is very simple:

$$Tp(\text{degrees}) = 360/N \qquad (4.4)$$

In units of time, Tp is equal to the actual spacing T divided by M:

$$Tp = T/M \qquad (4.5)$$

Equation (4.4) tells us that the effective spacing—*relative to one cycle of the signal*—can be made as small as we like, dependent only on N. This is the *informational* spacing.

Effective Sampling Rate

In coherent testing, it is the *effective* sampling rate that is governed by the Nyquist limit, not the actual rate. The effective rate, F_s', is the reciprocal of the primitive spacing:

$$F_s' = 1/Tp \qquad (4.6)$$

High speed sampling scopes and digitizers depend on this relationship to analyze GHz signals. In instruments that do not have the ability to unscramble vectors, F_s' is obtained by progressively delaying the sampling strobe by a time increment = Tp. This is simply a special case of a far more general and powerful consequence of M-over-N coherence:

$$F_s' = M/T \qquad (4.7)$$

$$F_s' = M * F_s \text{ (actual)} \qquad (4.8)$$

To provide an effective sampling rate of 10 GHz, for example, you could run the actual sampling clock near 10 MHz, and set the (PLL) synchronization to provide a ratio such as this one:

$$\frac{F_t}{F_s} = \frac{M}{N} = \frac{1001}{1024} \qquad (4.9)$$

The actual sampling rate F_s is 10000/1001 = 9.99+ MHz, and the test frequency is 10000/1024 = 9.7656 + MHz. The primitive or effective sample spacing. Tp is the reciprocal of 10 GHz, or 100 picoseconds.

Why would we need an effective rate of 10 GHz if the test signal is only 10 MHz? Perhaps the test signal is a super-sharp square wave, and we intend to look at DUT risetime, ringing, glitches, and other high frequency phenomena. This scheme allows us to look at *broad-band* information from DC up to $F_s'/2$ = 5 GHz, but permits an ADC rate under 10 MHz. With coherence and the unscrambling scheme shown later, Nyquist's limit applies not to the actual F_s, but to the effective rate F_s'. *The sampling system must have the same analog bandwidth and low jitter as if it were used at 10 Gs/s, however.*

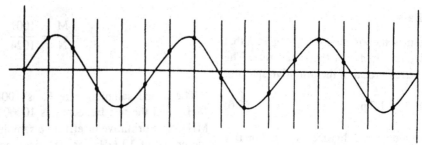

1. In coherent testing, the sample sequence repeats itself at regular intervals.
2. This interval is called the *primitive period*, and is an integral multiple of the waveform period.
3. It is the shortest valid test window, and thus often called the *unit* test period, or UTP.
4. The UTP contains N essentially *unique* samples.
5. The UTP contains M essentially *identical* signal cycles.
6. M and N are mutually prime integers.

(Above example: M = 3, N = 16)

Plate 4.1: The primitive period

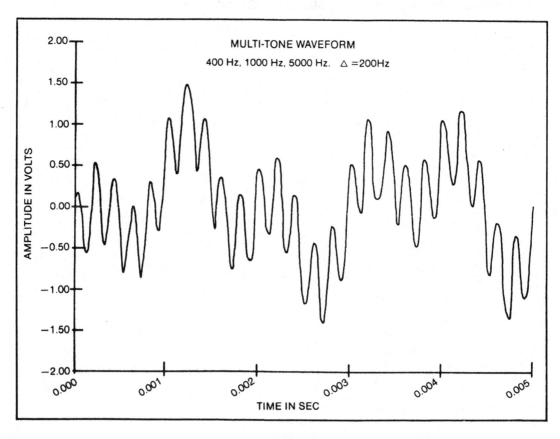

Plate 4.2: Multitone waveform

THE PRIMITIVE FREQUENCY

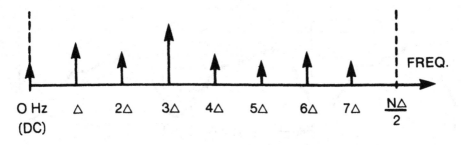

1. The sample set has its own frequency spectrum, not necessarily equal to that of the waveform.

2. This spectrum has N/2 + 1 discrete locations, or *"bins."*

3. Each frequency is a multiple of the *primitive* frequency, Δ.

4. $\Delta = 1/UTP$

5. All information is contained within the band, 0 to $N\Delta/2$, known as the *primitive band*.

6. All bins, except the two end bins, contain amplitude and phase information.

7. The frequency, $N\Delta/2$, is known as the Nyquist frequency. The Nyquist bin contains misleading information, and should not be used as a test frequency.

Plate 4.3: The primitive frequency

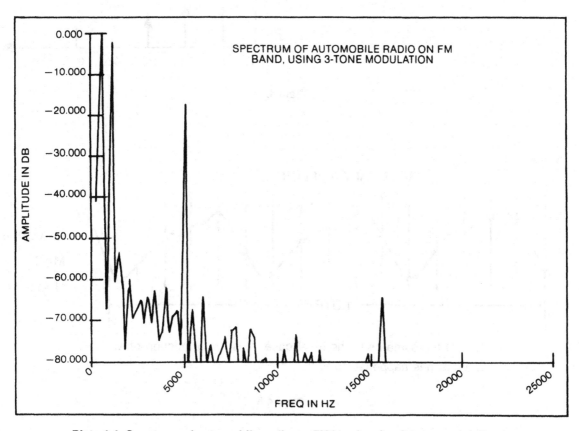

Plate 4.4: Spectrum of automobile radio on FM band, using 3-tone modulation

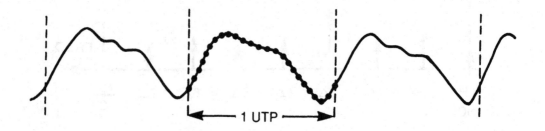

1 UTP

1. For maximum test speed, the UTP is set equal to the waveform period. M = 1.
2. N > 2M$_{(max)}$.
3. Explain what is the correct sampling frequency if M = 1?

4. Spectrum: When M = 1, the fundamental of the test waveform, F$_T$, equals the primitive frequency Δ. The spectrum is "normal."

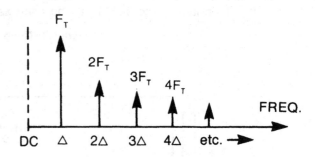

Plate 4.5

PRACTICAL SAMPLING : M > 1

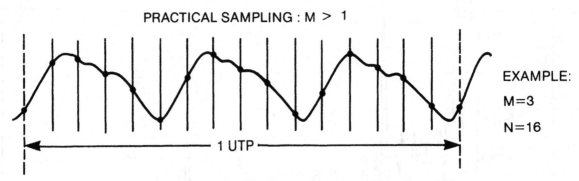

1 UTP

EXAMPLE:

M=3

N=16

1. N is the same as in the ideal case. N is not a function of M.
2. What has happened to F$_s$?

Plate 4.6

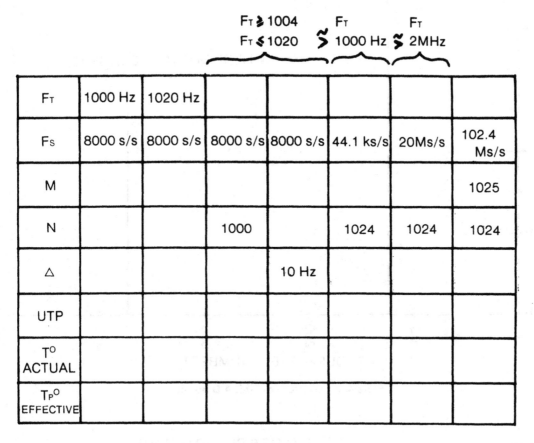

	Fт	Fт	Fт ≳ 1004 Fт ≲ 1020		Fт ≈ 1000 Hz		Fт ≈ 2MHz
F_T	1000 Hz	1020 Hz					
F_S	8000 s/s	8000 s/s	8000 s/s	8000 s/s	44.1 ks/s	20Ms/s	102.4 Ms/s
M							1025
N			1000		1024	1024	1024
$\triangle$				10 Hz			
UTP							
T^O ACTUAL							
T_P^O EFFECTIVE							

Plate 4.7

CASE STUDY: 6-BIT SPECTRAL TEST

(SEE FOLLOWING GRAPHS)

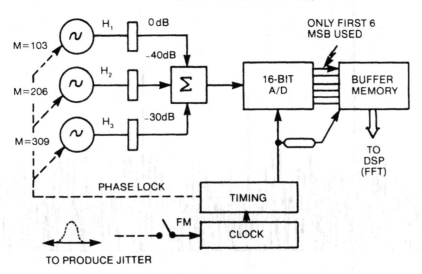

Plate 4.8: Case study of 6-bit spectral test

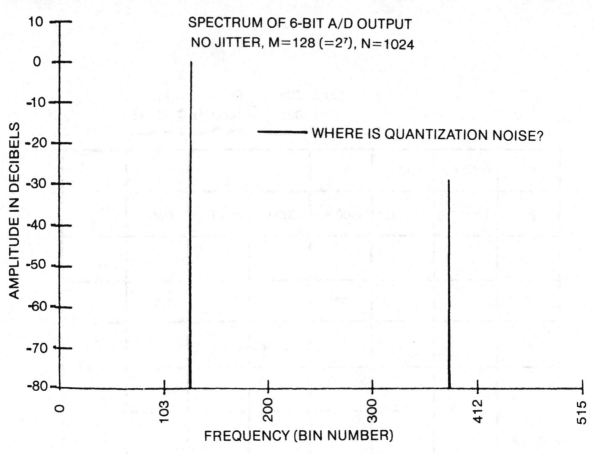

Plate 4.9: Spectrum of 8-bit A/D output

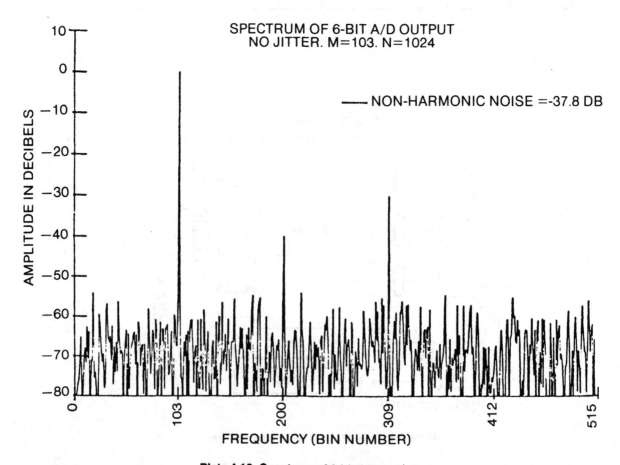

Plate 4.10: Spectrum of 6-bit A/D output

SPECTRAL SHUFFLING

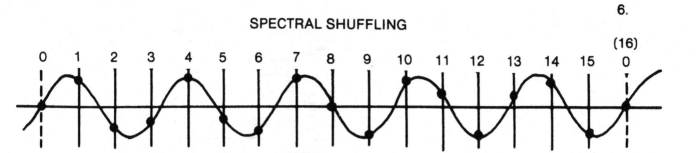

(16)

If the samples are processed in the sequence in which they are collected, the spectrum will be shuffled, step M, and reflected at the primitive boundaries.

EXAMPLE: M=5, N=16

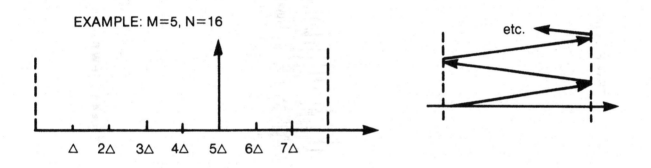

$\Delta \quad 2\Delta \quad 3\Delta \quad 4\Delta \quad 5\Delta \quad 6\Delta \quad 7\Delta$

etc.

WAVEFORM HARMONIC	DC	F_T	$2 F_T$	$3 F_T$	$4 F_T$	$5 F_T$	$6 F_T$	$7 F_T$
SPECTRAL LOCATION								

Plate 4.11: Spectral Shuffling

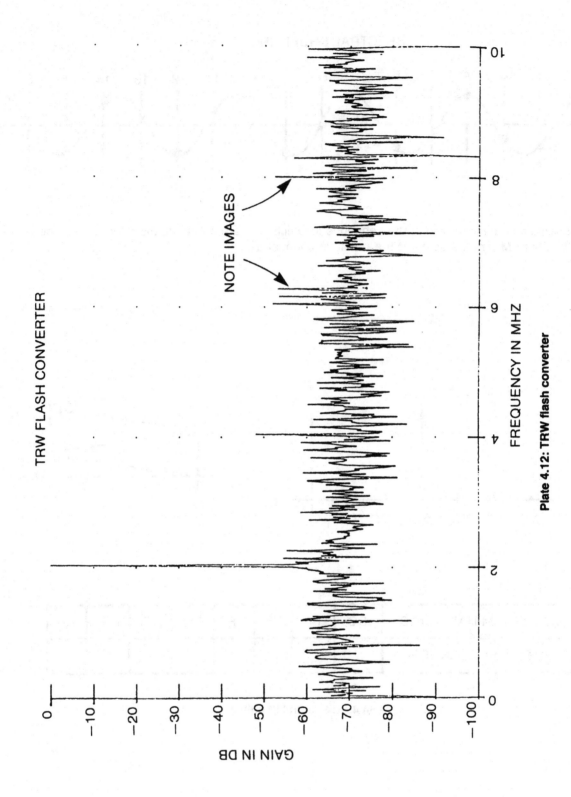

Plate 4.12: TRW flash converter

54

MODULO-N COUNTING

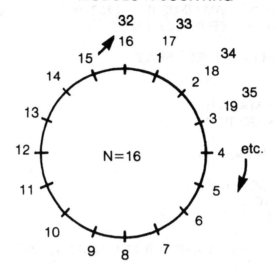

IN MODULO 16, 19 AND 3 ARE SAID TO BE CONGRUENT.

$$19 = 3, \text{ MOD } 16$$

EXERCISE: FIND INTEGER I SUCH THAT

$$I + M = 1, \text{ MOD } 16$$

WHEN M=5

Plate 4.13: Modulo-N counting

TIME SHUFFLING

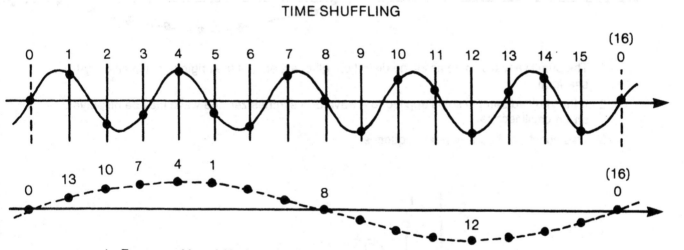

1. For every M and N, there is some integer I such that IM = 1, Mod N.
2. If the samples are processed step I, Mod N, they form the *primitive* wave.

Plate 4.14: Time shuffling

```
10      PRINT "THIS PROGRAM FINDS IS SUCH THAT M*I = 1, MODULO N"
20      PRINT / PRINT "ENTER INTEGER N"
30      INPUT N
40      PRINT "ENTER RELATIVELY PRIME M"
50      INPUT M
60      J = 0 \ I = 0
70      IF M > N THEN M = N - N
80      IF M > N THEN GO TO 70
90      J = J + M
100     IF J > = N THEN J = J - N
110     I = I + 1
120     IF I > N THEN GO TO 160
130     IF J < > 1 GO TO 90
140     PRINT "I = ", I
150     STOP
160     PRINT "★ ★ ★ ★ ★ ★ M AND N ARE NOT RELATIVELY PRIME ★ ★ ★ ★ ★ ★"
170     STOP
```

Plate 4.15: Program: MOD.MM

THE PRIMITIVE WAVE

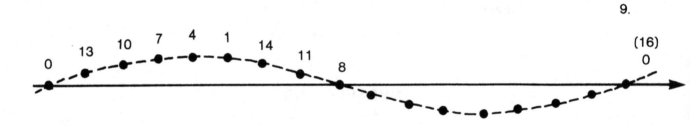

1. The primitive wave contains all the distortion, glitches, etc., of the original, but in an easy-to-see form.

2. If the shuffled sample set is applied to a waveform synthesizer, the wave may be displayed on an oscilloscope.

3. This spectrum has a normal sequence.

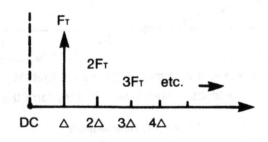

Plate 4.16: The primitive wave

SPECIAL CASES OF M/N SYNCHRONIZATION:
"BEAT FREQUENCY METHOD"

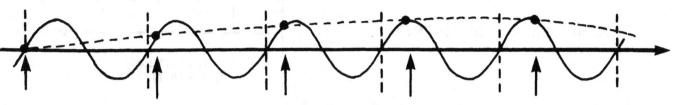

1. In coherent systems, the "beat-frequency" method is applied by setting M = N + 1, and I = 1.

$$\frac{F_T}{F_s} = \frac{N+1}{N}$$

2. Advantage:

3. Disadvantage:

Plate 4.17: Special cases of M/N synchronization: "Beat-frequency" method

SPECIAL CASES OF M/N SYNCHRONIZATION:

"ENVELOPE" METHOD

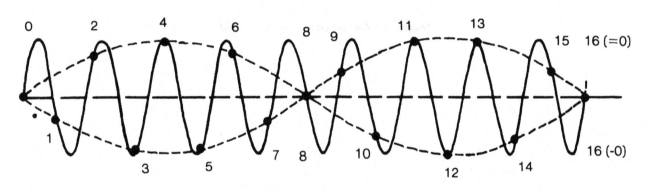

$$M = \frac{N}{2} + 1$$

WHERE N IS A MULTIPLE OF 4

1. Setting I = 1 (Direct Display) produces envelope pattern:

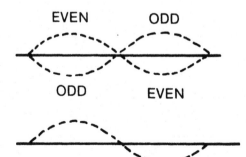

2. For primitive waveform, I = M:

Plate 4.18: Special cases of M/N synchronization: "Envelope" method

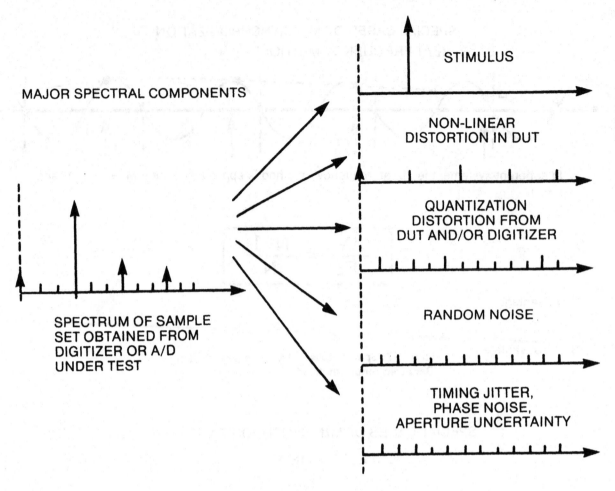

MAJOR SPECTRAL COMPONENTS

STIMULUS

NON-LINEAR
DISTORTION IN DUT

QUANTIZATION
DISTORTION FROM
DUT AND/OR DIGITIZER

RANDOM NOISE

TIMING JITTER,
PHASE NOISE,
APERTURE UNCERTAINTY

SPECTRUM OF SAMPLE
SET OBTAINED FROM
DIGITIZER OR A/D
UNDER TEST

Plate 4.19: Major spectral components

Chapter 5
Multitone Testing

Chapter 5: Multitone Testing

Coherent conditions make it possible to construct test stimulus signals composed of many mutually orthogonal sinusoidal tones. Orthogonal components are statistically independent, carry unique information, and can be separated without ambiguity. They may be simultaneously transmitted through the test device to produce a response vector from which many different parameters can be determined.

Consider the simple case in which the stimulus to an amplifier under test is the linear sum of two tones, 3f and 7f (Figure 5.1).

The period, P, is the computational window of the fast Fourier transform (FFT), and can be any whole multiple k of the unit test period (UTP). To introduce the concept, however, we will use the familiar condition where k = 1, and where the primitive frequency f equals the line spacing Δ. In this illustration, the device under test (DUT) input and the DUT output are digitized for 1 UTP each, while maintaining phase lock. Each vector is then converted to spectral form by the polar FFT. By comparing the two spectral vectors, we immediately obtain four parameters: the gain and phase shift at 3f and also at 7f.

(It is not necessary to digitize or analyze the DUT input for each new unit. The DUT input vector is the system *calibration* vector, which provides frequency and phase response information about the analog portions of the system synthesizer and digitizer, combined.)

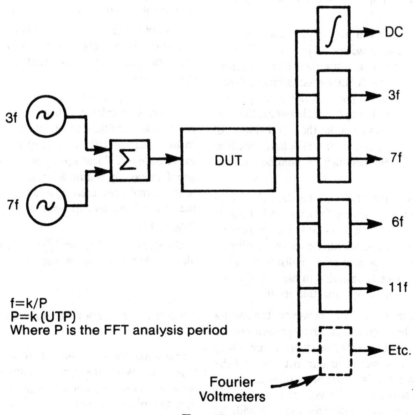

f=k/P
P=k (UTP)
Where P is the FFT analysis period

Fourier Voltmeters

Figure 5.1

Multitone Distortion Measurement

A similar procedure, using the magnitude FFT and three, four, or five tones, is often used as a fast screening test for low-, mid-, and high-frequency amplifier response. With properly chosen input tones, however, the response spectrum can provide a large amount of additional information. With the prime multiples, 3f and 7f, second harmonic distortion appears in bins 6 and 14. Third harmonics appear in bins 9 and 21.

With two or more tones, we can also examine intermodulation (IM) distortion. Second-order (sum-and-difference) components appear in bins 4 and 10. Third-order IM components fall in bins 1, 13, 11, and 17. Bin 0 contains the dynamic offset (i.e., the DC component with a signal applied). By using prime multiples for each input tone, the likelihood that two or more components will fall in the same bin is minimized.

A vector of N samples produces a spectral vector with $N/2 + 1$ bins. With a stimulus of two or three tones, this number is typically much larger than the number of significant harmonic and IM components. Many bins will consequently remain essentially unoccupied by these classical nonlinear distortion products. For simplicity, we will define them as "nonharmonic" bins.

Nonharmonic bins help us in two ways. First, quantization distortion and random noise spread out over all $N/2 + 1$ bins, making it possible to measure their effect independently of nonlinear distortion. A fairly good estimate of combined quantization and random power is obtained by measuring the average power per bin in the nonharmonic bins and multiplying by $(N/2 + 1)$. In addition, the fact that most of the nonharmonic power is removed from the harmonic bins improves the sensitivity and accuracy of nonlinear distortion measurement.

In A/D converter (ADC) testing, it is often desired to measure quantization distortion separately from the random noise. This is not practical when the test period P = 1 UTP, because the quantization errors are distributed over the UTP in a quasi-random pattern, producing a spectral distribution like that of random noise. This is desirable in measuring other DUT parameters, but not for quantization itself.

When P is a multiple of the UTP, however, the spectra are quite different. The quantization error pattern repeats from 1 UTP to the next, whereas random errors do not. There is a strong correlation between the unit patterns and the amount of correlation depends on the proportion of quantization distortion to random noise. By collecting a vector of 2 UTP, for example, we can correlate the first half with the second and, from the result, calculate the ratio of quantization distortion to random noise.

Separation can also be done in the frequency domain. Let the test period P equal k * UTP. The time vector will have kN samples, and the magnitude FFT will have $kN/2 + 1$ bins. The spectral line spacing Δ is no longer the primitive frequency, f, but a smaller value, f/k.

In the example of Figure 5.1., if there were no random errors, quantization distortion would fall only in bins whose numbers are multiples of k. Perfectly symmetrical quantization produces components at odd k only, while nonsymmetrical quantization produces components at all integer values of k. The latter is caused by DC offset or nonlinearity.

In the absence of random errors, bins that are not integer multiples of k are not occupied. If k = 4, for example, there are three empty bins for every one that contains quantization distortion. In this technique, *these bins are used to detect random noise.*

By adding the power in these bins, and multiplying by $k/(k - 1)$, we obtain an useful estimate of the total random noise of the converter under test, usually a combination of random voltage noise (Boltzmann noise) and jitter-induced sample errors. Adding the power of every kth line (only) gives an estimate of the quantization distortion. K should be large enough that the average power in the random bins is appreciably smaller than that in the quantization bins.

It is not necessary for tone *pairs* to be mutually prime in multitone testing. The requirement is that the *set* of all frequencies, *including* F_s, be irreducible (i.e., have no common factor).

Another application of making k greater than 1 is in minimizing interference from external frequencies upon a test whose conditions must not be changed. A test designed to use tones at 1020 and 1140 Hz, with $F_s = 8000$ s/s, experiences FFT leakage errors if subject to 50 Hz power interference. By setting K = 2, however, FFT bins occur at 10 Hz multiples, creating a valid bin for 50 Hz. This allows the original test conditions to be duplicated without incurring leakage.

Question: What is the UTP of 1020 Hz and 1140 Hz, alone? What is it when this pair is sampled at 8000 s/s?

Multitone Frequency Measurement

Multitone distortion measurements seldom use more than four tones. When the primary intent is to obtain frequency response, however, a much larger number of tones can be used to replace the usual "swept" frequency approach. The magnitude FFT provides amplitude versus frequency information, while the polar form adds phase versus frequency.

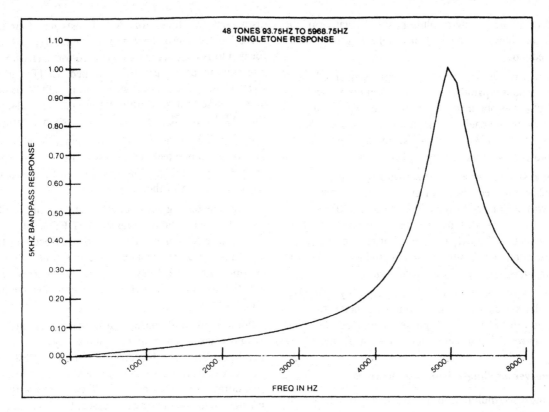

Figure 5.2

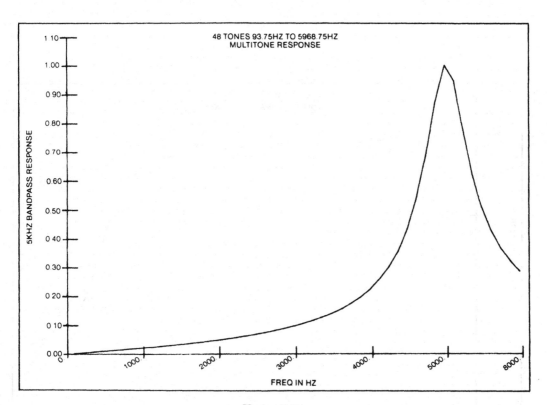

Figure 5.3

With sufficient frequency resolution, the derivative of the phase vector shows the amount of envelope (group) delay versus frequency.

A comparison of single tone response (sequential) with multitone response (parallel) is shown in Figures 5.2 and 5.3. The first plot was obtained by measuring the gain of a commercial switched-capacitor bandpass filter (National Semiconductor MF10) at 48 frequencies, one at a time. In the second plot, all 48 tones were applied at once.

The same synthesizer and digitizer were used in both tests, using 1024-point vectors. Both single tone and multitone analysis was performed by their FFT. In the multitone case, the stimulus vector was the numerical equivalent of 48 sinusoidal vectors, linearly summed. For reasons discussed later, these components were phased to produce a waveform with near-Gaussian amplitude distribution.

Figure 5.4 shows the phase-versus-frequency response. Relative phase accuracy for in-band tones is better than 0.1 degree (the limit of the bench phase meter used for reference). (Figures 5.2-5.4 were supplied by Jeff Stoyle.)

Multitone versus Single Tone Applications

For some tests, multitone testing is the only practical technique. By definition, a multitone stimulus is required for IM distortion, two tones for CCITT tests and four tones for AT&T telecom tests. Peak-to-average ratio (P/AR) tests performed to AT&T standards use a 16-tone set whose phases and amplitudes are precisely defined. CCITT noise-based CODEC tests use a multitone stimulus with 25 or more randomly phased lines. Multitone techniques are also required for DSP-based CCITT envelope or group delay tests.

DSP phase measurement can be made with either single tone or multitone techniques, but multitone testing is obviously faster and avoids the possibility of phase drift as the DUT warms up in the fixture.

Multitone testing does not automatically imply FFT analysis. The discrete Fourier transform (DFT) is needed in those cases where N cannot be a power of 2. Telecom tests involving trunk signaling are a case in point, where N must be a multiple of 6 or 12. In systems without an array processor, the DFT is also a relatively quick way to analyze just a few spectral bins.

For amplitude or phase measurement, there are many situations where a single tone stimulus may be preferred over a multitone stimulus. In a coherent sinusoidal test, for example, we can measure amplitude quickly and accurately in the time domain by means of the RMS operation or the absolute average. Without an array processor, this is much faster than either the DFT or the FFT and works with any N.

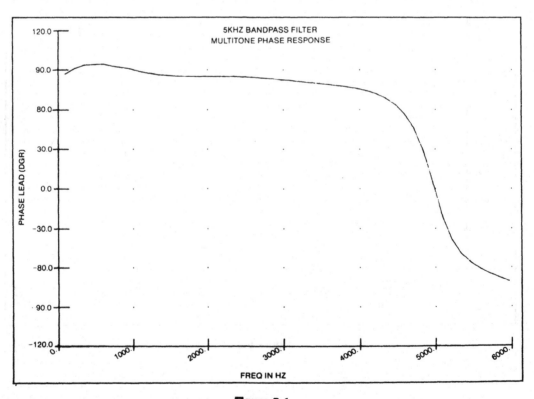

Figure 5.4

Although the root-mean-square (RMS) operation "sees" all spectral components, coherent RMS is a surprisingly good estimator of fundamental amplitude even with DUT distortion. The distortion is orthogonal, so its power, not voltage, is additive. With 1 percent harmonic distortion, the RMS voltage differs from the RMS of the fundamental (alone) by only 0.005 percent, or 0.0004 dB.

In measuring frequency response, multitone-FFT analysis has a clear advantage in speed. In the previous MF10 filter tests, the UTP was 32 ms, preceded by 15 ms of settling time. The FFT time plus computer overhead consumed 40 ms, giving a total multitone test time of roughly 87 ms. By comparison, the single tone sweep took 2.5 seconds even with an overlap of computation and digitizing.

Error Sources and Accuracy

Under otherwise equal conditions, multitone measurements are less accurate than single tone measurements. One reason is that the relative amplitude of the individual tones must be small to ensure that the peak-to-peak signal swing does not exceed the range of the DUT input. Each tone, therefore, experiences a poorer signal-to-noise ratio. For composites of four tones or more, the RMS uncertainty in the calculated amplitude of any individual tone increases as the square root of the total number of tones.

For devices that contain no quantization or digital filtering, however, multitone DSP accuracy is usually no less than that of conventional, single tone *analog* instruments. The reason is that DSP, of itself, gives enough head start over analog equipment to cancel out the degradation caused by multitone signals.

This category of devices includes the majority of amplifiers, active continuous filters, and switched-capacitor filters. In the previous 48-tone amplitude-versus-frequency test of the MF10 filter, the in-band uncertainty over 20 test runs was about 0.02 dB. By comparison, the uncertainty in 20 repeated single-tone sweeps using an all-analog analyzer (not DSP) was slightly larger, about 0.03 dB.

For maximum accuracy, single tone DSP is best. When the previous MF10 filter was swept by single tone DSP, the maximum variability over 20 runs was only ±0.003 dB.

Effect of Device Uncertainty on Multitone Tests

The test results from a batch of devices are always distributed or scattered in some way. The distribution can be characterized by its mean value, plus some measure of its scatter, typically, either the *standard* deviation, or the *peak* deviation. (With respect to the mean, the standard deviation equals the RMS value, and its square is called the *variance*. Variance is the statistical analogue of power.)

Ideally, the scatter is caused only by unit-to-unit manufacturing variation, or "production spread." The mean value is the "typical" device parameter. In practice, however, the mean value is shifted by calibration error, and the statistical scatter of measured parameters is greater than the production spread because of quantization effects and random noise of the test system and fixture and because of random noise (and possibly, quantization) within the DUT.

In this chapter, our interest lies with the *excess* scatter (i.e., that caused by dynamic uncertainties in the DUT and in the test system). We will use the term, "measurement uncertainty," to indicate the amount or range of excess variability in the computed value of a DUT parameter over a batch of measurements, beyond that directly caused by the production spread of that parameter.

In the days of all-analog circuits and instruments, nonrepeatability in testing the same unit again and again served as the measure of uncertainty. The major contributor was usually noise and/or unsynchronized interference, most of which typically originated in the test fixture or instruments.

Even if the source of nonrepeatability was noise within the DUT, measurement uncertainty was not lumped with the production spread. If large enough to be significant, it could be reduced by averaging several readings or by increasing the time constant of the detector or meter. This kind of thinking still dominates the test industry today, but, in fact, is not applicable in a great many test situations.

Modern digital signal processing (DSP) test systems and fixtures have considerably less random noise than their predecessors, and the same is generally true for most modern linear DUTs. If there is interference or crosstalk, it is more likely to be synchronous than random. The most significant difference, though, is a new form of measurement uncertainty based on discontinuous functions. Modern "linear" designs tend to be ones that lend themselves to very-large-scale integration (VLSI). They depend heavily on a mixture of discontinuous time (sampling) functions and discontinuous amplitude (quantization) functions (e.g., ADCs or D/A converters (DACs), comb filters, CODECs, digital filters, delta modulators, digital processors, and so on). Such circuits interact with test signals and measurement procedures in ways that can produce considerably more uncertainty than previous, all-linear circuits, and, generally, more than the digitizer of a modern DSP system. This uncertainty, when magnified by multitone techniques, can make good devices appear to fail.

The problem is that uncertainty introduced by synchronous interference and by discontinuous or nonlinear functions does not behave like that caused by random noise. In a coherent test system, it does not cause nonrepeatability in testing one unit, and so escapes the notice of an engineer looking for ''noise.''

In a coherent test, quantization errors repeat in every run, introducing a constant error. Averaging many readings does not improve the accuracy. You do not want to remove this error by a calibration constant, though, because the amount and direction of measurement uncertainty varies from unit to unit. The calibration factor that makes unit 1 behave better may make unit 2 behave worse. Over a large batch of measurements, measurement uncertainty may not alter the *mean*, just the scatter.

You may wonder why this type of unit-to-unit variability is not lumped with production spread. One reason is that the same variability can be seen in a single unit as the test conditions are varied; in other words, it is not simply a property of the batch. Another consideration is that, in principle, measurement uncertainty can be removed from any and all measurements with sufficient characterization and analysis.

To illustrate, suppose we are given a half-dozen reference mu-law encoders and are told that every unit has been carefully trimmed for unity gain, or 0.00 dB. For this batch, the production spread of the gain parameter is essentially zero.

We then apply a sinusoid of precisely known magnitude to the input of unit 1, decode the digital output sequence, apply the DFT or FFT to measure the fundamental component of F_t, and calculate the gain.

Instead of getting 0 dB, repeated tests of DUT 1 produce readings from 0.010 to 0.014 dB. Assuming that the variation of 0.004 dB is due to true random noise, the question is this: Is the *average* error of 0.012 dB a result of improper calibration of our system or of the system used by the person who trimmed the encoders?

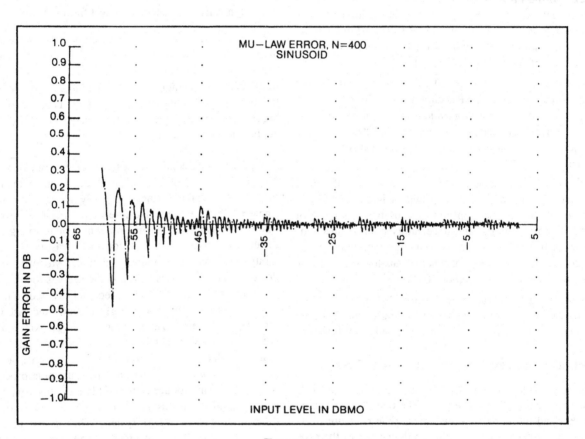

Figure 5.5

Neither. Both our system and that of the supplier might be perfectly calibrated. In fact, when we test unit 2, we obtain an error of about −0.09 dB, opposite to the previous error.

What seemed at first to be a calibration error is really measurement uncertainty in disguise. The way to unmask it is to vary the test conditions slightly and recalculate the gain each time. For example, we might advance the signal phase in 0.1 degree increments. In the days of all-linear circuits, this would have no effect on gain measurement. This time, however, a single unit produces a range of errors, in which errors of both polarities seem equally likely.

To experiment further, we vary the input signal level (amplitude) in small increments, say, 0.1 dB. This, too, causes the calculated gain to show a small and rapidly varying error. When we plot the results over a wide range of input level, we get the graph shown in Figure 5.5

The average of this curve is zero, showing that our system is actually calibrated correctly with respect to that of the supplier. We certainly cannot consider such variability to be part of the production spread, because it applies to a single unit.

If we could afford to test every unit this thoroughly, we could speak of quantization-induced *error*, rather than quantization-induced *uncertainty*. In a phase-locked coherent system, for example, repeatedly testing the same unit produces the same curve (except at low-input levels, where true random noise is proportionally large). From unit to unit, however, the curves will differ perceptively. The zero-crossings will differ enough that we cannot find a particular input level that will produce zero error, or even a predictable error, for *different* units of the batch. The true gain in any region is the average of the detailed curve in that region.

In summary, an error characteristic that is reasonably "certain" for a single unit—given thorough characterization—becomes uncertainty in production testing, where we must make brief measurements of many units. Making one detailed study for the entire batch has a great value, however, since it shows us the *range of uncertainty* to be expected in measuring the other units.

How does all of this concern multitone testing? Very often, devices are designed so that their intrinsic measurement uncertainty is small when compared with the production spread under the usual (i.e., single-tone) test conditions. Multitone testing magnifies this uncertainty, making the (apparent) production spread seem excessive.

Mu-law and A-law encoders with integral filters are examples of circuits for which multitone testing may not be sufficiently accurate. They have one of the worst cascade circuit

combinations from a test standpoint: a linear filter with tight specifications, followed by a low-resolution ADC. (A switched-capacitor filter, although discontinuous in time, is linear in voltage transfer.)

Multitone testing generally provides satisfactory accuracy for telecommunication filters that are separately testable. Typical manufacturing test limits for transmitting filter inband response run about ±0.07 dB to 0.10 dB, versus about ± 0.01 dB uncertainty for a 12-tone test.

When followed by an integral mu-law or A-law ADC, DUT uncertainty far exceeds that of the digitizer. As Figure 5.5. shows, mu-law encoding introduces about 0.01 to 0.03 dB peak single tone uncertainty at upper levels. A 12-tone multitone test can magnify this to the point that it is actually larger than the production spread. Even a perfect filter/encoder might appear to fail under these conditions.

Factors Affecting Accuracy

What is the error that should be expected in using multitone testing? The answer depends on a number of factors. Let us look first at four common problems, and then—assuming these are under control—give some simple formulas for the lower limits of uncertainty.

Tone Pruning

In creating a multitone stimulus, many of the tones that might otherwise seem desirable must be "pruned" or removed from the tone set to avoid conflicts among certain DUT distortion products. If you wanted to measure harmonic distortion of 3f and 7f tone inputs, for example, you would not want to include an input tone at 6f. This would obscure the identical, but much weaker, second harmonic of the 3f input component.

But suppose you do not care to measure distortion, only gain. And suppose the gain at 6f were the most important. Would it then be permissible to add 6f to the original set?

Possibly, since it is far stronger than the second harmonic of 3f. But the presence of any conflicting component will introduce uncertainty into the measurement.

The reader may recall the RMS example, where the presence of one percent distortion had no appreciable affect on coherent amplitude measurement. That was a single tone test, however, where the interference was from a different (and, therefore, orthogonal) frequency component. The present conflict, by contrast, is between two components of identical frequency, and their combined voltage may range from the difference to the sum. One percent distortion will introduce ± 1 percent uncertainty, or almost 0.09 dB. In short, if 6f is important, the input multiple at 3f should be pruned from the tone set.

There are a number of simple approaches used to assist in the pruning process. The rules vary according to the bandwidth of the tone set. This is often expressed in octaves (i.e., 2:1 frequency multiples).

For broadband sets of 2 or more octaves, a good start is to remove all even multiples. This prevents conflicts with second-order harmonic products and second-order IM products (sum-and-difference frequencies). The second step is to prune out any multiples divisible by 3.

Rather than pruning, some engineers prefer to choose multiples from a table of prime numbers. This obviously accomplishes the above results and eliminates certain other conflicts as well. This may remove one or more necessary tones from the set, however.

Question: Will using a prime tone set avoid third-order conflicts?

Narrow Band Tone Sets

Pruning is greatly simplified for narrow band multitone complexes. For those under 2 octaves in frequency span, using odd multiples is often sufficient; that is, there will usually be enough nonoverlapping bins to provide useful gain, distortion, and/or phase measurement.

In bands under one octave (e.g., CCITT noise-based testing) pruning is sometimes omitted completely (i.e., all lines are permitted).

Question: What distortion conflicts are avoided by tone sets less than 1 octave in bandwidth?

An indirect advantage of narrow band sets is that they contain fewer tones and thereby provide better signal-to-measurement noise for each tone. The disadvantage is that a broadband measurement cannot be made in a single test run.

Narrow band sets are sometimes very useful in testing filter response when both in-band and out-of-band response must

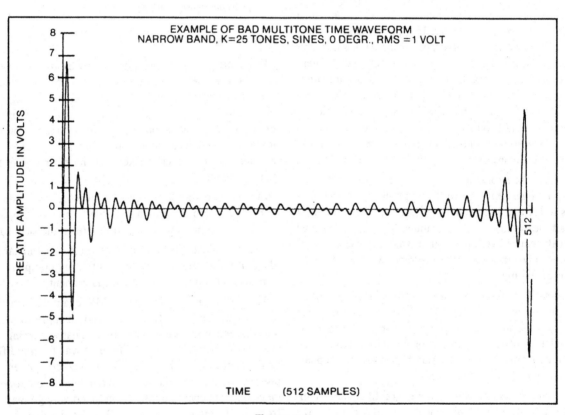

Figure 5.6

be measured accurately. By using sets that contain in-band tones (only) or out-of-band tones (only), we can adjust the gain of the DUT and/or digitizer for different sets to obtain the largest practical signal for digitizing.

Peak-to-RMS Ratio

When many tones are combined, the composite wave may have a high peak at one or more points. An obvious example is the situation in which all cosine waves are used; at t = 0, all the waves have an amplitude of +1. (In this tutorial, multitone waves are assumed to have equal amplitude tones unless otherwise noted.) A less obvious, but almost equally bad, example is the all-sine complex seen in Figure 5.6. Uniform phase distribution also produces a similar peaked wave.

When the amplitude of such a signal is adjusted to avoid DUT clipping, the remainder of the wave will be small when compared with the full-scale range of the device, and the effective measurement signal-to-noise ratio will be poor. To minimize the uncertainty of individual measurements, it is therefore desirable to use a test wave with low peak-to-RMS ratio.

If equally spaced tones are used, a very low peak-to-RMS ratio can be achieved by making the *differential phase* vary linearly from 0 to 360 degrees over the band of tone frequencies. The result is shown in Figure 5.7 for a one-octave tone set using odd multiples. If you look closely, you will note that it resembles a sinusoid with continuously swept frequency, changing at the end so as to loop smoothly back to the original pattern. (A sample BASIC program to generate this wave is given in the next chapter.)

Question: How does the wave (Figure 5.7) differ from one that is synthesized by periodically sweeping a constant-amplitude sinusoid up and down over a 2:1 frequency range?

At first glance, tones sets with minimum peak-to-RMS would seem to be ideal for multitone testing. In commercial testing, however, this is not the case because such waves are extremely sensitive to nonlinear phase in the DUT. As

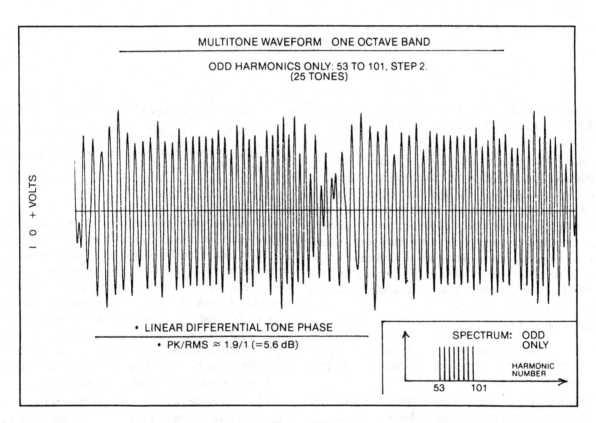

Figure 5.7

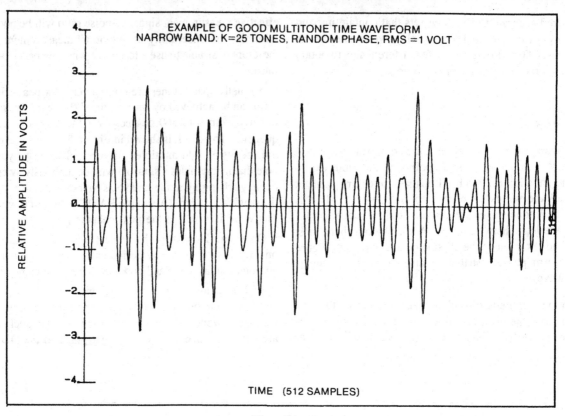

EXAMPLE OF GOOD MULTITONE TIME WAVEFORM
NARROW BAND: K=25 TONES, RANDOM PHASE, RMS =1 VOLT

RELATIVE AMPLITUDE IN VOLTS

TIME (512 SAMPLES)

Figure 5.8

they pass through the DUT, the peak grows with respect to the RMS value, and may overload the DUT at some internal point.

A clue to the optimum wave is given by the fact that by contrast, waves with high initial ratios tend to have smaller peaks after passing through the same real circuits. The most robust waves—the waveforms to which all multitone waves tend, with random or uncontrolled phase changes—are those with *near-Gaussian amplitude distribution* and pseudo-random phase distribution. The peak-to-RMS ratio will be about 10 to 11 decibels, a number that appears in many specifications without explanation. An example is shown in Figure 5.8.

Following are some exercises:

1. What is the theoretical peak-to-RMS ratio for 25 equal-amplitude cosine components? In decibels? Compare with Figure 5.6.

2. Does the RMS voltage of the multitone complex vary when the phase relationship is changed? Explain.

3. What is the peak-to-RMS ratio of a pure tone? In dB?

4. What is the peak-to-RMS ratio of the 2-tone set, (1,3)?

5. The primitive frequency is usually omitted from tone sets. Why?

Dynamic Overload

The maximum slope (maximum dv/dt) of multitone waves often exceeds that of ordinary test signals of equal peak value. For this reason, a multitone signal that appears to have a safe amplitude may still introduce excessive nonlinear distortion and increase the uncertainty of gain and phase measurements in active test devices, because of slew limiting, oscillation, or other dynamic effects. A simple way to tell if this is the cause of poor test accuracy is to try a somewhat smaller test signal. Often, reducing the input level just a few dB below what appears to be a safe level gives a dramatic improvement in measurement accuracy and repeatability.

Wide-band waves tend to have greater maximum slope values than narrow-band waves. As a consequence, multitone distortion tests often employ narrow-band stimuli, while wide-band signals are more likely to be used for amplitude-versus-frequency tests.

Out-of-Band Measurement Uncertainty

Multitone accuracy is often an issue in attempting to measure the complete response of a low-pass, high-pass, or band-pass filter with just one multitone wave. An example of this was seen in Figure 5.3.

The accuracy in measuring any individual tone out of a multitone complex is inversely proportional to the amplitude of that tone. In Figure 5.3, for example, tones above and below the pass band are attenuated by the filter under test. The vertical axis shows the actual ratio of output to input amplitude. At 3000 Hz, for example, a tone is attenuated to approximately 1/10th the strength of those in the pass band, and so experiences 10 times the measurement uncertainty of those in the pass band. The latter is about +/0.02 dB, so the 3 kHz tone has about 0.20 dB uncertainty. The tone at 500 Hz has about 1/20th of the in-band amplitude and so has an uncertainty of about 0.8 dB.

If accurate out-of-band response is required, it is best to use single tones for the key points of interest. This not only permits a larger input amplitude for each tone but also permits the digitizer input to be set to a more sensitive range. A compromise is to use separate multitone sets for in-band and out-of-band responses.

Estimating Multitone Accuracy

If the error source is predominately random noise, the uncertainty will be essentially the same as the nonrepeatability of a single test unit. The latter can be measured experimentally for a single tone and used to extrapolate the uncertainty for individual tones in a multitone complex. If, for example, a single tone experiences uncertainty U, then a tone component of relative amplitude 1/G will experience approximately G * U.

If U is expressed in decibels instead of percent, the same factor applies as long as the largest value is under 1 dB. For example, suppose that a strong tone experiences 0.10 percent uncertainty. A tone 1/5 the size is expected to exhibit about 0.5 percent. In decibels, U is 0.0087 dB for the strong tone and 0.043 dB for the weaker. When a calculator is not handy, you can estimate small percentage changes in amplitude by multiplying the number of dB by 11.

In discussing uncertainty, it is important to establish whether it is the RMS, peak, peak-to-peak, probable error,

or average error. In this tutorial, uncertainty expressed with the ± prefix refers to peak uncertainty; a number without it (unless otherwise noted) refers to the standard deviation, or "sigma." With reference to the mean of the distribution, this is the same as the RMS uncertainty.

If the uncertainty is a result of Gaussian (normally distributed) noise, and the mean is 0, then the various measures are related as follows:

1. RMS (= standard deviation):   1
2. Probable error: 0.6745
3. (Absolute) average error: 0.7979
4. Peak error (95 percent of samples):   ± 2
5. Peak error (99.7 percent of samples):   ± 3
6. Peak-to-peak error (common):   6

In an infinite vector, true Gaussian peaks are infinitely higher than the RMS. For N = 1024, however, only 3 samples are expected to exceed 3 sigma, and none is expected to exceed 4 sigma. On the average, only 1 sample out of roughly 20,000 will exceed 4 sigma.

Figure 5.9 shows the variation in 100 amplitude measurements of a steady tone because of the presence of random noise 40 dB smaller than the power of the complete 48-tone wave, or 48/10000 of the individual tone power. The expected RMS uncertainty is 10 log(1.0048), or 0.0208 dB. The observed uncertainty about the mean was 0.0212 dB.

Estimating Multitone Uncertainty Due to Quantization

The following formulas are based on the assumption that quantization distortion is the dominant source of measurement error and that the FFT is used to analyze the sampled response waveform.

Locate the dominant quantization source (DUT or digitizer) and express the integral nonlinearity error by the number of linear-equivalent bits. A good 15- or 16-bit ADC will usually have at least 14 bits of integral linearity. If the DUT uses nonlinear coding, or is dominated by noise, the number of *equivalent* bits can be estimated by expressing the signal-to-noise ratio in dB, then dividing by 6 dB. A mu-law CODEC, for example, has about 40 dB S/N, and therefore has about 6-1/2 bits of equivalent linear quantization error.

Use the number of equivalent bits to calculate the effective number of quanta in the full-scale range of the quantization source. Finally, express the peak-to-peak amplitude of the *individual tone* to be analyzed by the number of equiva-

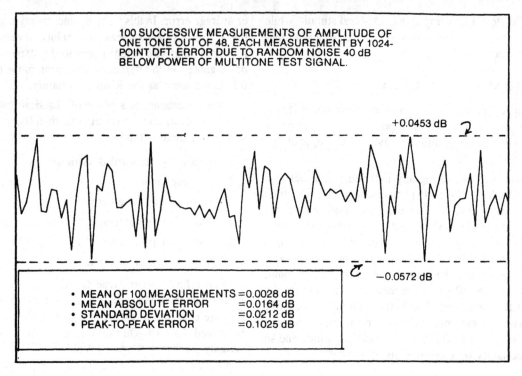

Figure 5.9

lent quanta, Nq. If the individual tone spans $1/10\ F_s$ of an (equivalent) 14-bit converter, for example, $Nq = 1/10$ of 16384, or about 1638.

1. Standard deviation in measured amplitude of a single sinusoid:

$$\sigma_s \approx \frac{1}{3\ Nq}\ dB \qquad (5.1)$$

2. Standard deviation in measured amplitude of 1 tone in a multitone complex (4 tones or more):

$$\sigma_m \approx \frac{1}{4\ Nq}\ dB \qquad (5.2)$$

3. Standard deviation in measured phase, 1 tone in a multitone complex:

$$\sigma_p \approx \frac{4}{Nq}\ degrees \qquad (5.3)$$

Note that quantization has slightly less effect on a tone that is part of a multitone complex, than on an isolated tone of equal amplitude. This is because the other tones in a multitone set act as "dither" signals, which move the samples pseudo-randomly with respect to A/D decision levels.

Figure 5.10 shows the error in 100 successive FFT amplitude measurements of a pure sinusoid caused by linear quantization where $Nq = 600$. To make the error vary over its range, the sinusoid was increased in amplitude 0.01 dB for each measurement. Figure 5.11 shows the uncertainty of the same amplitude tone, embedded in a 48-tone complex. Figure 5.12 shows the phase measurement uncertainty with 48 tones.

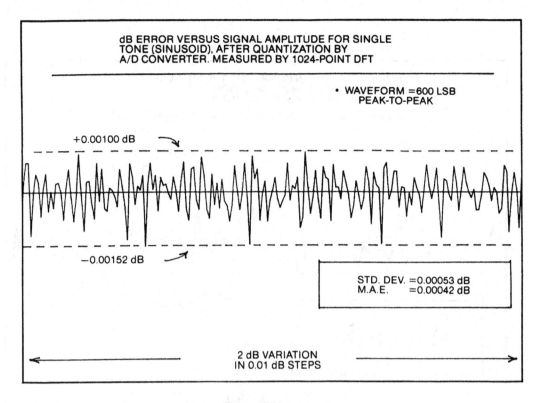

Figure 5.10

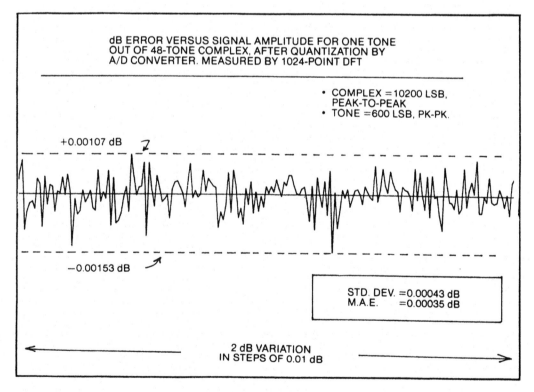

Figure 5.11

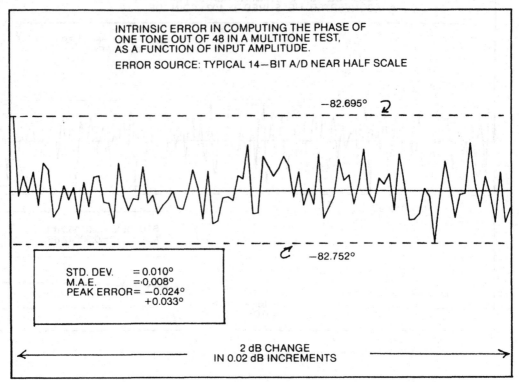

Figure 5.12

Chapter 6
Vector Operations for DSP Testing

Chapter 6: Vector Operations for DSP Testing

Analog hardware deals mostly with functions of time (waveforms) or functions of frequency (spectra). In DSP testing, however, the horizontal and vertical axes can represent anything the programmer wishes: leakage current versus temperature, distortion versus load impedance, and so on. In the most general sense, a DSP vector is simply the numerical equivalent of the classical continuous function, f(x). (See Figures 6.1 and 6.2.)

Since vectors can represent anything you wish, and since mathematical operations can do things instruments cannot,

digital signal processing (DSP) machines can do far more than emulate analog instruments. For the majority of production applications, however, DSP testing can be summed up in two brief statements:

1. Vectors represent waveforms, spectra, and filters.

2. Matrix and vector operations replace test hardware.

Here is a simple example. Given a distorted sinusoid D and an undistorted reference sinusoid U, where D and U are real electrical signals, we could subtract the two in a differ-

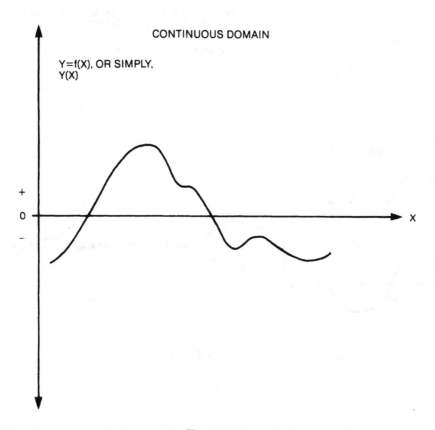

Figure 6.1

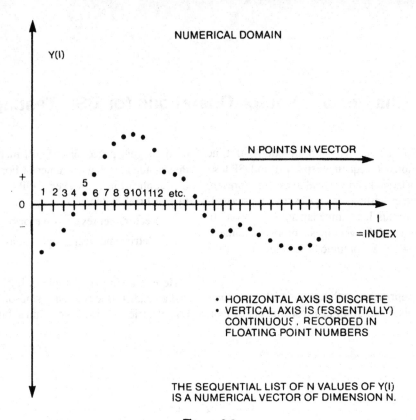

NUMERICAL DOMAIN

Y(I)

N POINTS IN VECTOR

1 2 3 4 5 6 7 8 9 10 11 12 etc.

0

+

−

I
=INDEX

- HORIZONTAL AXIS IS DISCRETE
- VERTICAL AXIS IS (ESSENTIALLY) CONTINUOUS, RECORDED IN FLOATING POINT NUMBERS

THE SEQUENTIAL LIST OF N VALUES OF Y(I) IS A NUMERICAL VECTOR OF DIMENSION N.

Figure 6.2

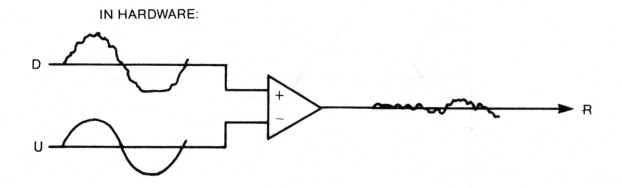

IN HARDWARE:

D

U

+

−

R

Figure 6.3

ence amplifier and display the remainder R on an oscilloscope. (See Figure 6.3.)

In a DSP system, by contrast, D would be obtained by digitization and temporarily stored as *vector* D. The undistorted version U would be determined by Fourier-filtering D. To find R, we would simply write the program line,

$$\text{MAT } R = D-U \qquad (6.1)$$

Not only is this simple, but all three vectors remain available for other tests. The noise voltage V, for example, can be found by the program line

$$\text{MAT } V = \text{RMS }(R) \qquad (6.2)$$

Vector Operations in DSP Programming

1. *Vector:* A vector is any list of N numbers to be processed as a related group. In analog testing, vectors usually represent waveform samples, spectral values, or filter coefficients. Vectors may be created within the·system computer, or obtained from the test device via the *digitizer or the receive memory.* Vectors may also be sent to the *transmit or send memory* to serve as stimulus patterns for a digital device or may be converted to analog waveforms via the *synthesizer.* (These four system elements were defined earlier.)

2. *Operation:* In DSP-based testing, the mathematical operation performed on a vector is usually a substitute for hardware. For example, the command,

$$\text{RMS }(X) \qquad (6.3)$$

simulates the behavior of an idealized RMS voltmeter. The vector X contains one or more periods of the input signal and the answer is the RMS voltage. The answer is not a vector in this example, but a single variable, or "scalar."

3. *MAT Notation:* When the operation involves one or more vector quantities, the BASIC prefix, MAT, is used to denote vector (or "matrix") arithmetic. Compare the following commands,

$$S = A + B \qquad (6.4a)$$

$$\text{MAT } S = A + B \qquad (6.4b)$$

Example 6.4a is the familiar *scalar* operation of addition; S is a single value. In 6.4b, the MAT prefix tells the compiler (and the programmer) that S, A, and B are each vectors of dimension $1 \times N$. Each element of S is the sum of the two corresponding elements in vectors A and B.

The MAT notation is not used in structured, non-BASIC languages. In structured DSP programming, variables S, A, and B would be defined in the program header as either scalar or vector variables.

4. *Integers:* Usually, each element of a vector is a 32-bit floating-point (FP) number, intended to convey a continuous range of voltage. The letters I, J, K, L, M, and N, however, denote 16-bit integers, positive whole numbers from 0 to 65,535. Thus the command,

$$\text{MAT } K = S \qquad (6.5)$$

converts each of the FP numbers in vector S to the nearest available whole number.

This particular operation, by the way, directly represents the action of an ideal 16-bit unipolar A/D converter (ADC). One LSB equals 1 unit of relative input voltage.

As noted above, the MAT notation is not used in non-BASIC languages. Some syntaxes allow the programmer to define vectors as either integer or FP in the program header, while some employ prefixes or other identifiers.

5. *Logic Operations:* In programming, integer vectors like J, K, and L are dimensioned as simple $1 \times N$ matrices. From a logic viewpoint, such vectors are actually $16 \times N$-bit Boolean matrices. If N is, say, 4096 words, then the logic operation,

$$\text{MAT } L = J \text{ XOR } K \qquad (6.6)$$

produces an answer of 16×4096, or 65,536 bits, where each bit is the exclusive-OR of the corresponding bits in matrices J and K.

6. *Constants:* The bracket notation, [ ], indicates that the enclosed quantity is applied equally to all *numerical* values of a vector. To set all elements of X to 1.026, for example, we write,

$$\text{MAT } X = [1.026] \qquad (6.7)$$

To add a bias of 1.026 to each element of waveform W, we could write,

$$\text{MAT } W = [1.026] + W \tag{6.8}$$

7. *Indexed operations:* Any vector operation can be restricted to a desired subset of elements. To find the sum of the elements of X from the 31st to the 220th position, we would write,

$$\text{MAT } S = \text{SUM } (X \ (31 \ \text{to} \ 220)) \tag{6.9}$$

BASIC has an advantage over certain other languages for DSP in this aspect. Indexing is a standard feature of the pre-defined data structure. Limited indexing such as illustrated in (6.9) is especially useful in computing the power in different spectral bands.

8. *Building-Block Concept:* For maximum versatility, basic engineering functions like filter, rectifiers, and amplifiers, are represented by separate mathematical commands. Frequently encountered "building-blocks" are shown in Figure 6.4.

Program Examples

Multitone Wave with Linear Phase Difference

In the chapter on multitone testing, it was stated that a waveform with low peak-to-RMS ratio is produced when the tone-to-tone phase difference linearly increases from 0 to 360 degrees over the band of tone frequencies. The following program shows how the waveform vector might be formed in a BASIC derived DSP language. In this example, F and T represent the frequency (spectral) vector and time (waveform) vector, respectively. Line 10, not shown, dimensions these to the chosen N. Scalar P is the tone phase, and D is the difference between adjacent phases.

The frequency vector follows a common format in which the DC component is placed at index 1 and the Nyquist component, index 2. The primitive frequency pair (cosine, sine) occupy positions 3 and 4, and so on. In general, the cosine and sine parts of the Jth multiple occupy index positions $(2J + 1)$ and $(2J + 2)$.

```
20    P = 0
30    D = 0
40    MAT F = [0]
50    FOR J = 53 TO 101 STEP 2
60    F (2 * J + 1) = COS (P)
70    F (2 * J + 2) = SIN (P)
80    D = D + 15
90    P = P + D
100   NEXT J
110   MAT T = INVERSE FFT (F)
```

In this example, there are 25 tones spanning less than 1 octave (53 to 101 Δ, odd multiples only). Since there are 24 phase increments, the increment D is increased by 360/24, or 15 degrees in each execution of the loop. The resulting wave was plotted in Figure 5.7.

Multitone Wave with Random Phase

A multitone wave with near-Gaussian amplitude distribution can be produced by a similar program using random phase.

```
200   FOR J = 53 TO 101 STEP 2
210   R = 360 * RND (X)
220   F (2 * J + 1) = COS (R)
230   F (2 * J + 2) = SIN (R)
240   NEXT J
250   MAT T = INVERSE FFT (F)
```

The scalar operation "RND(X)" produces random numbers between 0 and 1, with uniform distribution.

Broadband Gaussian Noise Vector

```
300   FOR I = 1 TO N
310   R1 = RND(X) + 0.00001
320   R2 = 360 * RND(X)
330   T(I) = SQR(-2 * LN(R1)) * COS(R2)
340   NEXT I
```

This routine converts the uniform distribution of the BASIC operation "RND(X)" into a normal (Gaussian) distribution with a standard deviation of 1.000. In line 310, an arbitrarily small number is added to R1 to prevent the possibility of R1 being zero. The natural logarithm operation (LN) cannot accept an argument of zero.

Ideal n-Bit Quantization

Let X represent a floating-point waveform vector and Y represent the same waveform after passing through an ideal quantizing channel (A/D-D/A) of n-bits resolution. M is an intermediate (scratchpad) vector composed of integers only.

The procedure is first to scale the amplitude of X so that its range is slightly less than the full scale range (FSR) of the converter to be simulated. For example, if n equals 12 bits, then the range of X should be less than $\pm$ 2048, if bipolar, or less than 4096, if unipolar. If the DSP operating system cannot accept negative integers, a constant K is added to X to create a unipolar intermediate vector.

```
400   MAT M =[K] + X
410   MAT Y = [-K] + M
```

HARDWARE FUNCTION	TYPICAL SOFTWARE EQUIV	NOTES
1. INTEGRATOR 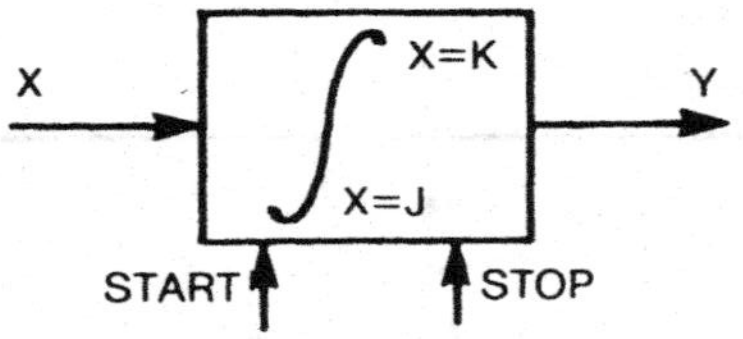	MAT Y=SUM (X(J TO K)) OR, MAT Y=SUM (X)	SCALAR ANSWER
2. DC VOLTMETER 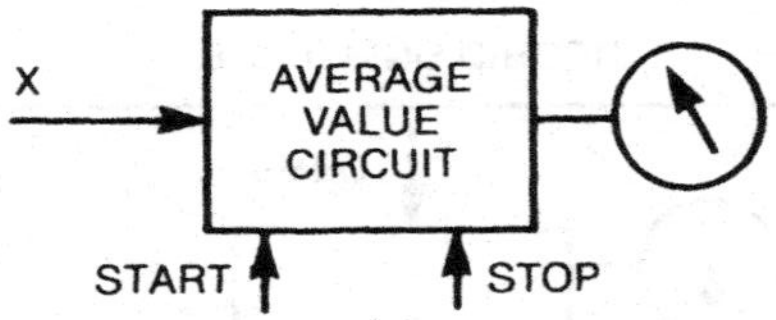 AVERAGE VALUE CIRCUIT	MAT Y=AVG (X)	SCALAR ANSWER
3. FULL-WAVE RECTIFIER 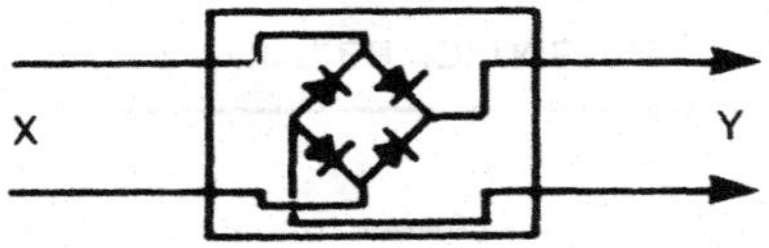	MAT Y=ABS (X)	VECTOR ANSWER: DIM (Y)=N
4. CODEC A/D, D/A 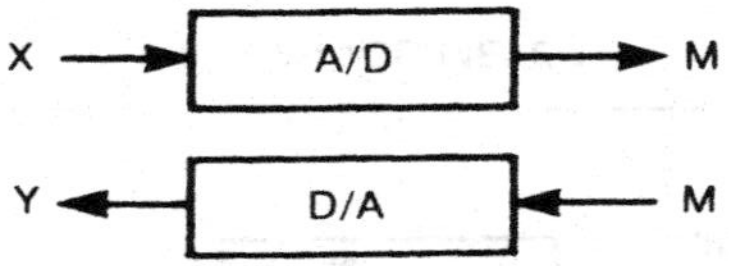	MAT M=MUCODE (X) MAT M=ACODE (X) MAT Y=MUDEC (M) MAT Y=ADEC (M)	

Figure 6.4

HARDWARE FUNCTION	TYPICAL SOFTWARE EQUIV.	NOTES
5. PEAK CATCHER	MAT Y=XTRM (X)	DIM (Y)=4; ANSWERS INCLUDE INDEX.
6. RMS VOLTMETER	MAT Y=RMS (X)	IF AVG (X)=0 THE RMS OPERATION GIVES THE STANDARD DEVIATION
7. FOURIER VOLTMETER	Y=FVM (X, J, N) OR MAT Y=DFT (X)	2F
8. CLIPPING CIRCUIT	MAT Y=X UBOUND BY V	COMPLEMENTARY OPERATION IS LOWER BOUND: MAT Y=X LBOUND BY [V]

Figure 6.4 continued

HARDWARE FUNCTION	TYPICAL SOFTWARE EQUIV.	NOTES
9. SPECTRUM ANALYZER 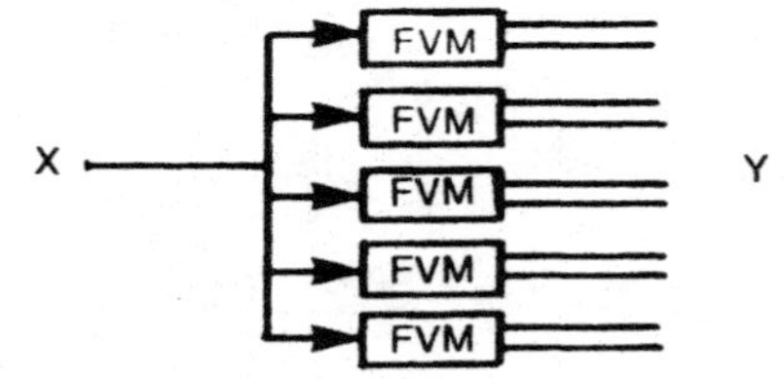	MAT Y=FFT (X) MAT Z=POLAR (Y)	DIM (Y)=DIM (X)=N=2'
10. MULTITONE SYNTHESIZER 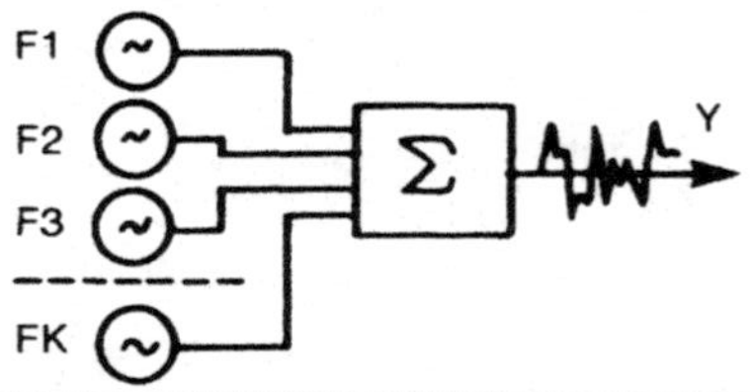	MAT Y=INVERSE FFT (X)	DIM (Y)=DIM (X)=N=2'
11. UNIVERSAL FILTER 	MAT Y=X FILTERED WITH R	DIM (Y)=DIM (X)+DIM(R)−1 FOR OPEN FORM DIM (Y)=DIM (X) FOR CLOSED (PERIODIC) FORM
12. CORRELATOR 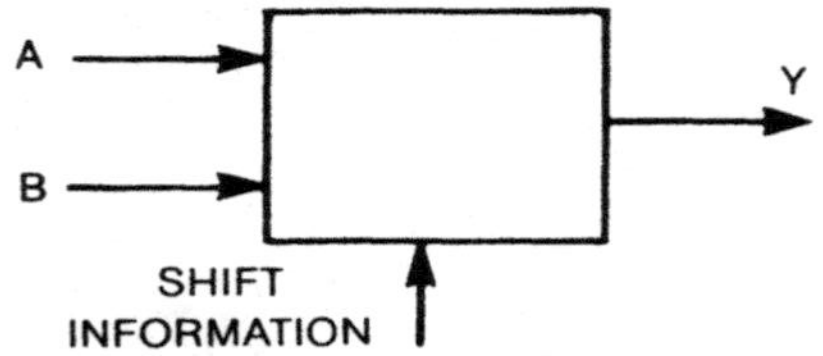	MAT Y=A CORRELATED WITH B −J TO K	DIM (Y)=J+K+1

Figure 6.4 continued

Chapter 7
Event Digitizing

Chapter 7:   Event Digitizing

The thing that distinguishes digital signal processing (DSP) from other forms of computer mathematics is that it deals with numerical vectors instead of scalars. To put it another way, a DSP system allows the programmer to mathematically process entire strings of informationally related numbers as simply as single variables in a hand calculator.

A numerical vector is the sampled equivalent of some continuous function. If the continuous function is y(x), then the sampled version might be denoted Y(I), where I is the index integer, = 1, 2, 3, ... (Figures 7.1 and 7.2).

The function itself can be anything the test engineer wishes: voltage versus time, amplitude versus frequency, offset versus temperature, and so on. The vector stores only the values of the function y(x) giving no information about x. It is up to the programmer to keep track of the values of x to which the set of y(x) correspond.

The computer keeps track of the successive samples by the index I. Element Y(35), for example, refers to the value of y in the 35th position from the left, *not* the value of y sampled at x = 35. If vector Y represents the frequency response of some filter, starting at 50 Hz, and continuing in 10 Hz increments, then Y(35) is the gain at 390 Hz.

There is a tendency to equate the term "discrete" with "integer," because the horizontal (x) values are commonly located by indexing. But discrete simply means that the choice of numbers is limited, with obvious gaps between them. These numbers can have as much resolution as you like, or even be irrational. Their spacing need not be regular or orderly (although that is usually the case when x represents time). In the broadest sense, index integers merely form a set that can be mapped into another, noninteger set of numbers X(I).

In Figure 7.2 only the horizontal axis is intentionally discrete. Even though the vertical axis is quantized by finite-length coding, it is effectively a continuous axis because all the code words are usable (i.e., the gaps are much smaller then the gaps along the horizontal axis). This is most important, for as long as one axis is effectively continuous, it is still possible to place all samples squarely on the curve of the original continuous function.

What may be a surprise is that even though one axis is intentionally quantized by a limited number of points—which means that the vector is blind to most of the function—the continuous curve can nonetheless be *fully defined in all regions* by a properly placed discrete set of samples.

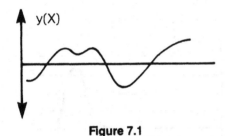

Figure 7.1

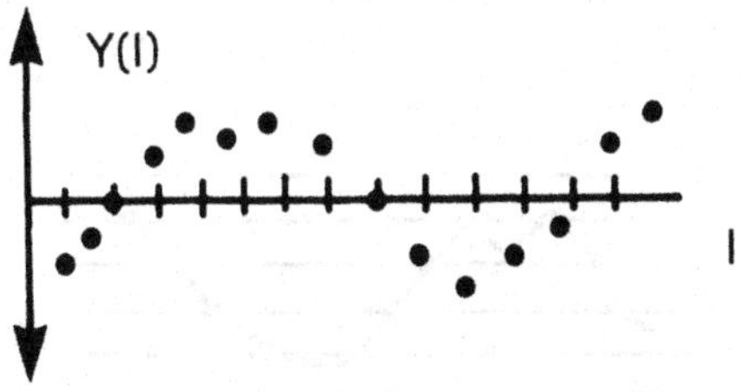

Figure 7.2

This statement is true regardless of which axis is made discrete. The only difference is that there are different rules for proper placement. However, it seems to be an unwritten law in nearly every DSP text, nearly every sampling scope and nearly every DSP-based instrument, that the numbers to be displayed or processed relate to the vertical axis.

This unwritten law reflects the fact that numerical rules for properly placed samples are extremely simple in the case of uniform time distribution and are highly successful. But it should not blind us to the possibility of making the vertical axis discrete and recording continuous values from the *horizontal* axis. It is this scheme that we are now going to investigate.

Explicit versus Implicit Digitizing

Conventional sampling can be viewed as a process in which the sample points are intersections of the continuous curve of y(x) with a set of vertical lines (Figure 7.3). In the new scheme, we superimpose a set of horizontal lines instead (Figure 7.4).

Algebraically, this new vector is said to provide *implicit* information about y(x). Instead of recording response values that have been forced by selected values of the independent variable, the vector records the values of the independent variable x, which are required to force selected values of the dependent, or output, variable y.

You may recall from calculus that it was sometimes easier to integrate or differentiate certain equations implicitly rather than explicitly. Suppose you want the derivative dy/dx but are given an equation in the form x = g(y). Yes, you could solve first for y, then differentiate, the explicit solution. But for some equations, it will take less paperwork to differentiate the equation as it stands. Well, something similar often occurs in DSP problems, and it is very handy if our DSP-based test system has the ability to digitize and synthesize signals either way.

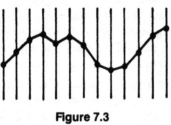

Figure 7.3

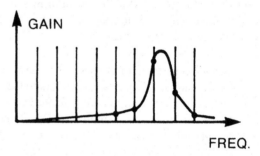

Figure 7.5

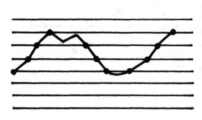

Figure 7.4

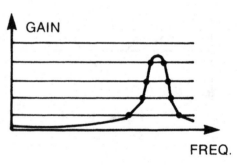

Figure 7.6

There is more to the story, though. The great majority of functions we are likely to encounter in automatic test equipment (ATE) applications are single-valued functions: For every stimulus point, or x value, there is one and only one corresponding response or y(x) value. The filter response curve in Figure 7.5 is such an example. A sine wave is another. Every vertical line is crossed once and only once.

In the implicit scheme of Figure 7.6, however, you see that some horizontal lines never intersect the curve, while others are crossed more than once. There is no one-to-one "point mapping" between the two vectors. This offers the possibility that a very small vector of one kind may be just as useful as a very long vector of the other kind, and will greatly increase test speed.

Suppose we need to find the center frequency of the above filter, somewhere near 1 kHz. We might also like to know the peak gain and the 3 dB frequencies. In the usual ATE single tone approach, the test frequency F_t would be stepped through the region around 1 kHz, and the output amplitude repeatedly measured. Multitone DSP would permit all the frequencies to be applied at once, but those frequencies would still be ones that are arbitrarily selected. Unfortunately, none of these provide a simple answer to the problem (Figure 7.5).

It is unlikely that any of the preselected x values (frequency) will fall exactly at the 3 dB gain points or at the maximum gain point. In this scheme, the answers must be interpolated from the entries in the vector. Since it is also unlikely that the samples will fall symmetrically with respect to the peak, accurate interpolation involves some curve fitting.

Now contrast this approach with a hypothetical system offering implicit measurement. The output (vertical) axis is discrete and the frequency axis is essentially continuous. Only a few horizontal lines are needed, placed near the expected peak output. Exact placement is not important. Analysis can be restricted to the highest line intersected. This line either has one value of x (in which case x happens to be the center frequency) or two values (in which case the center lies midway between the two values (Figure 7.6).

Once the center frequency is known, the old vertical-line concept is better for finding the peak gain: Just set the input frequency to that value and record the output.

However, having found this, we return to the horizontal-line concept to locate the 3 dB frequencies. Just calculate the level 3 dB below the peak gain and place a line at that level. This single line provides two solutions (i.e., both of the points of interest).

In general, any problem where the horizontal location of key features must be determined is a job for *implicit* digitizing. *Explicit* digitizing is aimed at finding the vertical location of key features that correspond to predetermined inputs.

To contrast the two, consider the following two ways in which amplifier distortion might be tested. In the first, we are asked to determine the distortion at 1 watt output. In the second, we are to determine the output power at 1 percent distortion. The first is an explicit (vertical-line) test. The second is an implicit (horizontal-line) test.

To measure distortion at 1 W, the gain of the device under test (DUT) is first measured by using a small signal. From this gain, the ATE computer calculates and applies the exact input for 1 W output, then measures the distortion.

The second test is not so straightforward for conventional test systems. An input is applied and distortion is measured. If the distortion is less than 1 percent, the input is increased, and a new distortion measurement is made. The process is repeated until the distortion falls within an acceptable zone around 1 percent. If, at any time, the distortion exceeds this zone, the succeeding input level is reduced. When the distortion is close enough to 1 percent, the output power is measured. The number of steps can be reduced by interpolating between points, but distortion is a nonlinear function of signal level, and linear interpolation would not be accurate unless the points were reasonably close to one another.

Although both tests are equally easy to state in words, the second is far more cumbersome and time consuming. This is because we have tried to make an implicit measurement on a test system designed only for explicit measurement.

Until recently, virtually all DSP systems were focused on those types of problems where the explicit, or vertical-line, approach was used: They applied predetermined x values, then measured y(x). This is the reason why linear ATE systems have traditionally been poor at automated measurements of rise time, fall time, propagation delay, overshoot, settling time, glitch size and energy, wow and flutter, and timing jitter. These are things that call for high resolution measurements along the *horizontal* axis (time) of events that are defined by the intersections of the waveform with just a few horizontal lines. What we intend to do now is show how the system structure can be altered to facilitate implicit measurement.

Event Digitizer

In new DSP test systems, the physical process that approximates implicit measurement is called *event* digitizing. In event digitizers, as with conventional digitizers, the horizon-

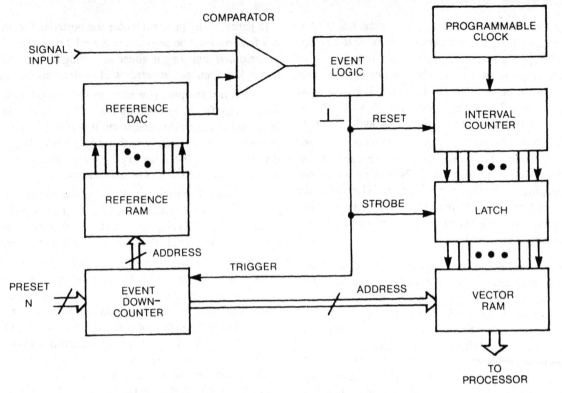

Figure 7.7: Single-channel event digitizer

tal axis is time. But instead of recording successive samples of voltage, the event digitizer records essentially continuous values of *time* at which predetermined "events" occur along the vertical axis. The time axis is not regularly sampled, but is recorded with high resolution when and only when an output "event" occurs.

An "event" is the crossing of the vertical-axis parameter with a horizontal threshold. The y parameter is usually the voltage output of the device, or something easy to convert to voltage, such as the derivative of voltage, or current. For DUT distortion, y could be the voltage output of a real-time (hardware) Fourier voltmeter (FVM), and the input would be a sinusoid with progressively increasing amplitude, x. This would permit y(t) to bear a known relationship to y(x).

The event digitizer does not use an analog-to-digital converter (ADC) because the voltage y(t) is not measured. In implicit digitizing, voltage is a preset quantity, produced by a digital-to-analog converter (DAC). Waveform intersections are determined by high-speed comparators. It is the relative *time* of each event that is digitized, and the vector is composed of hundreds or thousands of time measurements (Figure 7.7).

Suppose we wish to place hundreds of horizontal lines across the waveform to be analyzed. Does this mean we need hundreds of comparators and hundreds of reference DACs. Not at all. No real band-limited analog wave can cross two different levels in the same instant of time. It is thus possible to use a few high-speed comparators whose DAC threshold voltage is altered "on the fly." That is the function of the reference RAM in the single-channel event digitizer of Figure 7.7.

There is a limit to how fast a single DAC/comparator can settle to a new state, so practical event digitizers usually employ two or more such channels, connected to trigger in some sequence. One pair, matched for propagation delay, can be fired sequentially for things like rise time. For many measurements, it is desirable to program one pair to trigger on up-going crossings only and another pair on down-going crossings. A fifth DAC/comparator is useful in providing a trigger to open a time window for certain events.

The idea of using a comparator, a reference DAC, and a counter is an old one, standard in ATE timers and frequency meters. The event digitizer, however, is designed to produce *vectors* that contain many thousands of time measurements. There must be a vector RAM, for instance, capable of storing one or more complete vectors. Another RAM can store the sequence of thresholds is stored in advance, plus instructions to alter this sequence on the fly, according to DUT behavior, without intervention by the computer.

The *interval counter* is always reset to zero at each event, quickly enough so that none of the subsequent clock pulses is missed. When the next event happens, the interval count is latched, the counter is reset, and the process repeats. These latched counts are placed into sequential addresses of the vector RAM. The vector thus stores only time *differences*, not absolute time, which means that the number of bits in successive elements does not accumulate. 16-bit words are typical.

With 16-bit resolution, each interval of time between two successive events can be resolved into 65535 parts, more than adequate for high-speed intervals like rise time and propagation delay. Far greater resolution is available for repeated events, however, by averaging many elements. A vector of length 1000, for example, can contain over 65 million clock counts.

There is another type of counter in the event digitizer: the *event counter*. This is preset to the desired number of intervals, and decremented every time an interval count is latched. It stops at zero and provides an interrupt signal to the DSP computer. The vector length is thus controlled by the programmer, not in time, but as a specified number of events. A timeout circuit is provided in case an event does not occur within a programmed time limit.

Testing Tape Decks

There are many commercial ATE applications of event digitizing. One of the simplest, yet one that nicely illustrates the power of the technique, involves the production testing of tape decks. For this, only one horizontal threshold is required, held constant throughout the test. The comparator is programmed to detect up-going zero-crossings only, which means we will measure the cycle-by-cycle periods of the input waveform.

In this example, the following parameters are to be determined:

1. Frequency response at 50, 315, 3150, 6300, 10000, and 12500 Hz
2. Harmonic distortion at 315 Hz
3. Noise level
4. Tape speed
5. Peak wow and flutter (peak frequency variation)
6. DIN (weighted) wow and flutter

The example is drawn from a real program for a standard automotive cassette radio. To reduce test time, a multitone tape is used that contains all six "tones," or sinusoidal frequency components, at once. That is, each track of the tape contains a complex waveform made up of six components, locked coherently so that they share a common period, the unit test period (UTP). These frequencies are IEC standards.

If this test were of an amplifier, where the output frequencies remain constant, conventional waveform (discrete time) digitizers could do the job efficiently: A band-split filter could feed everything below 1000 Hz (the two lowest frequencies and second and third harmonics) to one digitizer and the remaining four to the other.

The two low components are multiples of 5 Hz (Δ), where M_i are 10 and 63, respectively. The numer of samples collected in the vector N has to be twice the highest M. Assuming no appreciable distortion above the third harmonic, M (max) is 189, so 512 is sufficient for N. This is a power of 2 and permits the use of the fast Fourier transform (FFT) for rapid spectral analysis of gain, distortion, and noise for that part of the spectrum. The sampling frequency F_s is 512 * 5, = 2560 Hz. The UTP is 200 ms.

The high components are multiples of 50 Hz, giving M_i of 63, 126, 200, and 250. The set has no common divisor. This particular automotive deck rolls off sharply after 12500 Hz, so N = 1024 is more than adequate. F_s is 1024 * 50 Hz, or 51200 Hz. The UTP is 20 ms.

If these tones are reproduced by a mechanical tape deck, the above test cannot be done accurately unless some additional DSP procedures are used. One of the things that complicates the test is that the absolute frequencies will be shifted by some unknown amount, different for every unit. One of the tests, you recall, is to determine the amount of speed error. Second, the tape speed fluctuates with time because of vibration and out-of-round idler wheels, capstan, etc. In this particular example, the combination of variations introduces frequency modulation with frequency components down to about 1/3 Hz. In other words, we want to record the tape output for 3 seconds or more in order to make sure we see the extremes of speed.

In itself, 3 seconds is no problem. It is still relatively short when compared with non-DSP methods that may take a minute or more by using a separate cassette for each tone and by using non-DSP instruments. For conventional (i.e., discrete time-axis) digitizing, however, there are simply too many samples to collect and process. The low-rate unit would have to collect more than 2560 * 3 samples, at least 8000. This is larger than the maximum FFT size for many array processors. The high-rate unit is worse: over 3 * 51200, or over 153600 samples, totally impractical for commercial DSP.

There is more trouble. DFTs and FFTs are accurate only if they operate on a whole number of signal cycles. But a conventional digitizer has no way of knowing when a specified number of *cycles* has been captured, just the number of samples. Yes, there are techniques for dealing with fractions of cycles, but they are not highly accurate in this situation, and they further complicate the DSP calculations.

A better way to solve the problem is the use of an event digitizer in addition to the two conventional waveform digitizers. The event digitizer is used first to measure the speed and frequency variations (things that are characterized by horizontal uncertainty), and its information is then used to set F_s to match the average frequency. In practice, even though the waveform digitizers are not coherent, this procedure will make the frequencies sufficiently precise that the FFT or DFT will give better amplitude and distortion measurements than traditional analog meters.

Speed measurement does not involve the waveform digitizers at all, and can be accomplished very accurately by event digitizing. The information in the event vector also gives a detailed picture of the speed variations.

For speed measurement, the 315 Hz tone is separated from the others by a band-pass filter and sent to the event digitizer. The filter has a flat top sufficiently wide to accept the production spread of tone frequencies, plus the associated FM sidebands.

The event digitizer collects only *one* sample per cycle: the interval of each cycle. In the following example, the event counter is set to collect exactly 1024 samples, regardless of the rate at which they occur. If the actual playback frequency is between 300 and 330 Hz, the test period will be between 3.1 and 3.4 seconds.

The vector thus formed is transferred from the RAM into address K.

$$\text{MAT READ} < \text{address of RAM} > \text{INTO K} \qquad (7.1)$$

MAT K is a list of 1024 cycle widths. The first computation is one that gives the *average* width:

$$\text{MAT A} = \text{AVG(K)} \qquad (7.2)$$

The first element of A is the average of all 1024 cycle-by-cycle counts, expressed in clock counts. In the commercial application from which this example is taken, the clock is set at 5 MHz, giving 200 ns increments. If the tape speed were exact, each cycle at 315 Hz would take 3174603.2 nanoseconds, or 15873 clock counts.

Suppose, however, that A(1) = 15777 for the tape deck under test. This means the tape speed is low by 96 counts out of 15873, or 0.605 percent. Note how simply this was obtained. As a thought experiment, you may wish to determine how the same result could have been obtained by conventional explicit (waveform) digitizing and DSP.

Next, let's look at the "wow and flutter" (i.e., the speed variations). This subject is discussed at length in Chapter 11, which shows how DSP is made to conform to formal standards (ANSI, DIN, CCIR, etc.). Here, our primary concern is to show only how the event digitizer provides the information needed for speed variation analysis, and we will restrict the algorithms to a few simple types.

To examine the speed variations, first subtract the average A(1) from the original event vector:

$$\text{MAT E} = [-A(1)] + K \qquad (7.3)$$

MAT E is the *error vector*, a list of cycle-by-cycle deviations from the average period. To find the root-mean-square (RMS) error, for example, write

$$\text{MAT R} = \text{RMS (E)} \qquad (7.4)$$

To find the peaks, or extremes, of error, write

$$\text{MAT X} = \text{XTRM (E)} \qquad (7.5)$$

Or, to find the (absolute) average, write

$$\text{MAT A2} = \text{ABS (E)}$$
$$\qquad (7.6)$$
$$\text{MAT A3} = \text{AVG (A2)}$$

Remember, the vector MAT E is not a signal or a list of sampled voltages. It has nothing to do with amplitude. Nevertheless, we can still take the Fourier transform if all we want to see is the frequency distribution of the time errors.

Although the error vector has the exact number of points and tone cycles (1024) for an accurate transform, it also has wow and flutter components that are not periodic over the

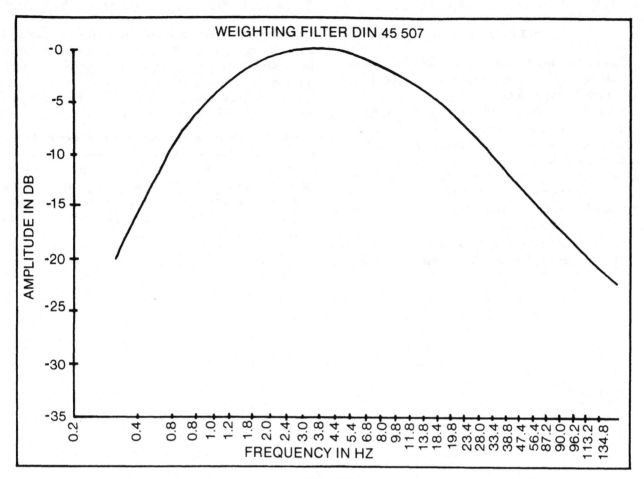

Figure 7.8

test period. To minimize the error, the vector is first multiplied by a "window" that squeezes the ends to zero, forcing the series to be periodic over the vector interval, that is, to loop without a discontinuity. A raised cosine (the window of Von Hann, or simply, a Hann window) is sufficient for this task and is easy to compute. It is here labeled W and stored for use as needed.

$$\text{MAT E2} = \text{W} * \text{E} \qquad (7.7)$$

Now the FFT may be applied:

$$\text{MAT F} = \text{FFT (E2)} \qquad (7.8)$$

The result, MAT F is the spectral plot of the speed variation errors, or wow and flutter. It should not be confused with the FFT obtained from waveform vectors.

Since the subjective intensity of flutter varies with its rate, the spectrum can be weighted by the Deutsche Industrienorm (DIN) standard weighting curve, which peaks at the frequency of maximum human sensitivity to frequency variations in sound, close to 4 Hz (Figure 7.8).

The program includes a precomputed vector that represents the frequency response of this filter. If we label it MAT D, then we may write

$$\text{MAT E3} = \text{D} * \text{F} \qquad (7.9)$$

MAT E3 is the *frequency weighted* spectrum of the speed variations of this tape deck. A typical logarithmic plot is shown in Figure 7.9.

To find the spectrally weighted "power," take the sum of products of the weighted spectral lines:

This answer, like the others, is in units of clock counts. It must be normalized to a fraction of tape speed by the appropriate conversion factor.

Figure 7.10 shows the amplitude response of the four high tones for the same automotive tape deck used above. The broadening of the lines at their base is a result of leakage from noncoherent conditions. The tones are far enough apart, though, that the leakage does not appreciably alter the amplitudes of the dominant lines. This leakage is far less than if the sample rate F_s had simply been set according to the *nominal* frequencies.

Unlike the event digitizer, the waveform digitizers need work only over the short test periods calculated earlier for spectral measurement only: 200 ms for the low band and 20 ms for the high band. The frequency variation broadens the spectral lines, but a good answer is obtained by summing the power in the spectral regions at each center frequency. The component at 12500 Hz is weak because of poor head azimuth alignment in this unit.

Actually, the UTP for waveform digitizing is so short when compared with the 3 seconds required for the wow and flutter test, that only one waveform digitizer was used, first for the low group and then for the high group. The total test time for all the parameters listed earlier, including all computation, transfer, and overhead, was slightly more than 5 seconds.

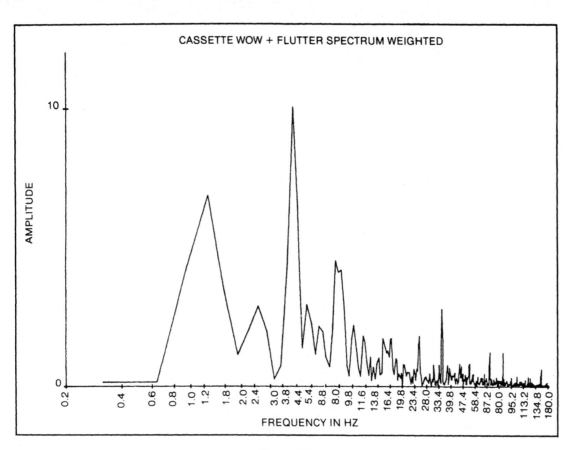

Figure 7.9

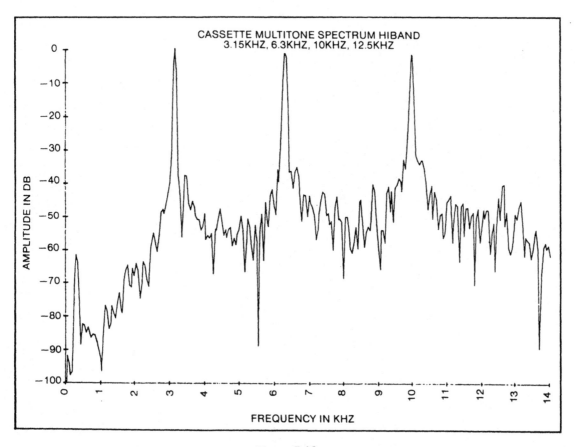

Figure 7.10

Chapter 8
Measuring Random Noise

Chapter 8: Measuring Random Noise

Understanding the technique of noise measurement is not so much a matter of knowing digital signal processing (DSP)—which is relatively easy in this case—but a matter of understanding the statistical nature of noise and being able to relate this to certain properties of electrical circuits. The measurement of ''op amp'' noise and spectral noise density provides a good illustration.

Equivalent Input Noise

Noise is generated by every component in an amplifier. It propagates forward and is amplified along the way. The composite noise is measured at the output terminals, usually through a bandpass filter that restricts the frequency range to a specified region of interest. In a DSP system, the band limited output noise is digitized, and the resulting vector is analyzed mathematically.

Although noise is measured at the output of the device under test (DUT), it is usually specified at the input, as *equivalent input noise* or as noise ''referred to input'' (RTI). The reason for this is that the amount of output noise voltage depends not only on the ''noisiness'' of the internal circuitry but upon the amplification of the DUT. To provide a universal measure of comparison, noise voltage is *normalized* by dividing the measured output noise by the amplification or gain of the DUT.

Equivalent input noise has two distinct components: voltage and current. Voltage noise is the combined effect of internal components and by definition is the noise that reaches the output when the input terminals are dynamically short circuited. The equivalent input voltage is expressed by a model that has the same gain as the amplifier under test but is free of all noise except for an internal noise voltage generator (zero internal impedance) placed in series with the upper input terminal.

Current noise, by contrast, physically exists at the DUT input terminals. It is produced by the junctions and related components in the input path and can be represented in the above model by placing a noise current generator (infinite internal impedance) across the two DUT input terminals. The current cannot flow into the DUT model, because the input impedance of the model amp is infinite. It flows instead through the signal source or feedback network, producing a noise voltage proportional to the driving point impedance. This noise voltage is in series with, but independent of, the equivalent noise voltage generator.

To evaluate *current* noise, we use an input network with high impedance, so that the developed voltage (current times resistance) dominates the internal voltage noise. To measure the *voltage* noise, a low impedance is used. To illustrate, the OP-07 amplifier typically exhibits about 0.35 microvolts voltage noise in the 1 to 10 Hz band and about 14 picoamperes current noise (both p-p). The fixture's source impedance should thus be either much greater than, or much less than, 25 K ohms.

Since the combined voltage is the RMS sum of the individual voltages, the ratio does not have to be large for one component to dominate the other. If the ratio is only 5:1, for example, the measured noise is only 2 percent greater than that of the larger component alone. This error is less than the typical variation of repeated measurements of random noise.

Figure 8.1 shows a simplified fixture for measuring voltage noise. The DUT is placed in a feedback network that sets the gain at a value repeatable from unit to unit (i.e., a value far below the open-loop gain) yet high enough (e.g., 1000 x) to amplify the voltage noise well above that of the test fixture. A typical value of R1 is 100 ohms.

In Figure 8.1, amplifier A1 is the DUT and A2 is part of the fixture. If the DUT noise were 1 microvolt RTI, setting G1 to 1000 would produce an output noise level of only 1 millivolt. This is not high enough to be transmitted reliably through the long cables of a large test system, so additional amplification (G2) is provided by A2.

A1 amplifies its own DC offset as well as its noise. For most op amps, the offset voltage is much greater than the noise voltage, and the amplified value could easily saturate A2. In this fixture, a coupling capacitor is used to block the amplified offset of A1.

Normalized Spectral Noise Density

The amount of noise measured by the digitizer depends not only on the DUT and the gain but also on the noise band-

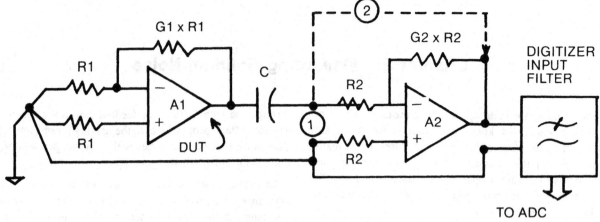

Figure 8.1: Noise fixture

width. To provide a further universal measure of comparison, the equivalent input noise is also normalized by expressing the noise that would (in theory) be measured by using a 1 Hz bandpass filter. The result is said to indicate the *noise density* at the center frequency, and has the dimensions of volts per root hertz. The working unit is most often nanovolts/root Hz. Sometimes noise density is expressed as the noise in a bandwidth of 1 Hz.

Spectral density is often confusing to new test engineers, because it seems to imply that we must be able to measure nanovolts and must use a filter with 1 Hz bandwidth. "Root hertz" is also an unfamiliar concept. But spectral density is only a calculated parameter. Just as we do not actually measure input noise at the input, neither do we measure nanovolts or root Hz nor use a 1 Hz bandpass filter.

On the contrary, it is desirable to use the widest practical measurement band, because measurement uncertainty is inversely proportional to the square root of the measurement bandwidth. By using 400 Hz for the measurement band, the repeatability is improved by a factor of 20 compared to that obtained with 1 Hz bandwidth. Alternately, for same uncertainty, test time can be reduced by a factor of 20.

To measure the voltage noise density at 1000 Hz, for example, the actual measurement band might be 800 to 1200 Hz. In the DSP approach, we measure the relative noise *power* in this band and determine the power-to-bandwidth ratio. The dimensions are volts squared per hertz, or (across a hypothetical 1 ohm load) watts per hertz. For ideal (white) noise,

this is an important ratio because it is independent of frequency or bandwidth and thus an *intrinsic* parameter of the DUT; it is the noise power density. The square root gives the more familiar parameter, noise voltage density, in volts per root Hz.

For white noise, the nominal measurement frequency f_n is the arithmetic mean of the lower and upper limits f_1 and f_2. In semiconductor amplifiers, however, noise density tends to increase with decreasing frequency, and it may be necessary to adjust f_1 and f_2 so that the average power density equals the expected density at f_n. For pink noise (-3 dB per octave, amplitude, or 1/f power), f_n is the *logarithmic* mean of f_1 and f_2. For brown noise (-6 dB per octave, or 1/f amplitude), f_n is the geometric mean:

$$\text{logarithmic } f_n = \frac{f_2 - f_1}{\ln(f_2) - \ln(f_1)} \qquad (8.1)$$

$$\text{geometric } fn = \sqrt{f_1 * f_2} \qquad (8.2)$$

For bands of 1/2 octave or less, the *arithmetic* mean generally suffices for all distributions. If the limits are 80 and 120 Hz, for instance, f_n is 100 Hz for the arithmetic mean, 98.65 Hz for the logarithmic, and 97.98 Hz for the geometric. The difference is insignificant in noise testing.

If the band were 20 to 180 Hz, however, f_n would be 72.8 Hz for pink noise and 60 Hz for brown noise. In this case, it would be best to characterize the distribution in this region and to determine f_1 and f_2 that most accurately represent f_n.

Typical DSP Procedure

DSP noise measurement begins with digitization of the broadband noise from A2 (e.g., the noise from 1 to 1400 Hz). The noise vector thus formed is transferred to the DSP computer. The first DSP operation is one that sets the vector *average* to zero.

Applying the root-mean-square (RMS) function to this zero-mean vector gives the standard deviation of the broadband noise, which is also the RMS noise voltage. To find noise density in various smaller (local) bands, the fast Fourier transform (FFT) is used to obtain the power spectrum. From this, the power in local bands is computed by the sum of squares (SSQ) operation. The local bands might be 800 to 1200 Hz, 80 to 120 Hz, and 3 to 33 Hz (for f_n = 10 Hz with 1/f noise). The vector power *density* in each region, normalized to 1 ohm, is computed by dividing the power by the bandwidth in hertz. The square root gives the noise voltage density in volts per root hertz and is normalized to the DUT input by dividing by (G1 * G2).

Here is a sample DSP BASIC program:

Instruction	Remarks
MAT READ WD INTO K	"WD" is a waveform digitizer. K is the noise vector: a set of ADC code words
MAT A = AVG (K)	A(1) is code offset + DC (fixture) offset
MAT X = [−A(1)] + K	X is the "AC" noise vector
MAT R = RMS (X)	R(1) is broadband RMS noise in LSBs
R = R(1) * C	Constant C = RTI nV per LSB; R is thus broadband input noise in nanovolts.
MAT F = FFT(X)	F is complex noise spectrum
MAT P = SSQ (F(L1 TO L2))	P (1) is peak power in band defined by limits (indices) L1, L2. (Peak power, obtained by squaring sinusoidal amplitude, is twice the average or true power.)
V = SQR (P(1)/W/2) * C	W is the measurement bandwidth in Hz; P(1)/W/2 = power density; V is noise voltage density, RTI, in nanovolts per root Hz

The broadband region is defined jointly by the digitizer filter and the FFT operation. The filter cutoff frequency establishes the upper limit, roughly equal to or less than half of the sampling rate. But often this filter is flat down to DC.

In that case, the lowest frequency of analysis is the primitive frequency (i.e., the reciprocal of the test period). If F_s is 3072 s/s, for example, the filter cutoff Fc might be set to 1400 or 1500 Hz. If 2048 samples are collected, the test period = 666.6 ... ms, and the primitive frequency is 1.5 Hz. The range of 1.5 to 1400 Hz would be appropriate for the measurement bands around 10, 100, and 1000 Hz. For noise at f_n = 0.1 Hz, F_s and Fc would have to be reduced by a factor of 100 to achieve the same uncertainty as at 10 Hz with the first set of conditions.

Input Resistors

In measuring voltage noise, the input resistance value R1 must be kept small not only to minimize current noise error but also to minimize the voltage noise that these two resistors contribute by themselves.

Thermal agitation of electrical charges within the resistors creates an independent, random local current in each incremental element dR of a resistor. The resistor thus acts as a series string of independent random voltage generators, and the total RMS voltage is proportional to the square root of the total resistance.

In an ideal resistor, the noise voltage (the so-called Boltzmann noise) is uniformly distributed across the frequency spectrum, meaning that its spectral density is independent of frequency. At 300 degrees K, the RMS noise density Dn in nanovolts per root hertz, is approximately

$$D_n \text{ (RMS)} \approx 4\sqrt{R(k\Omega)} \qquad (8.3)$$

Real resistors make more noise than this, but good film or noninductive wire wound units should not be much worse, say, half again as much. Carbon composition resistors are often far noisier than ideal.

Exercise 1: Assume that R1 (Figure 8.1) is 100 ohms and that the resistor pair generates twice the ideal Dn. How much input noise does the pair contribute in a 4 kHz band?

Exercise 2: How does this affect the noise measurement of an amp with 4 nV RMS/root Hz RTI?

Coupling Capacitor

Generally, the DC offset voltage of the DUT is much greater than the noise voltage, too large to be amplified by both the DUT and the preamplifier. In this example, a coupling capacitor C is used to prevent the amplified DUT offset from saturating the preamplifier.

Ideally, the time constant, C * R2, should be greater than the reciprocal of the lowest frequency component to be measured. This causes the settling time to be much larger than the desired test time, however, and production fixtures usually include some means of momentarily reducing the time constant for rapid settling. Closing a relay or field-effect transistor (FET) switch in positions 1 or 2 for a few milliseconds will do this without degrading fixture performance (Figure 8.1).

Other methods are used to speed up settling that do not use a capacitor but are usually no faster in practice than the fast-charge scheme shown here. Remember, the amount and polarity of offset varies from unit to unit and is unknown prior to the test. It may also drift as the DUT warms up during test. For drift, the capacitor is better than injecting fixed DC to oppose the DUT offset.

Noise Bandwidth

In DSP-based noise testing, the digitizer input filter usually is set to the full band of interest and the RMS operation is used to calculate the noise voltage over in this band. This is the broadband noise. While this is not always part of the device specifications, it *always* should be measured in debugging the test plan. The combined gain G1 * G2 should be large enough that this broadband noise will span a substantial portion of the digitizer's input range, but never so large that the noise is clipped. In selecting gains G1 and G2, we should consider three factors:

1. The *broadband* noise at the digitizer input, *not* the density
2. The peak-to-peak noise, not RMS
3. The noise of the worst-case acceptable device

Note that the bandwidth to be considered is not defined by the −3 dB (half-power) points but by an imaginary "brick wall" (infinitely sharp) filter whose power-frequency area equals the integral, over frequency, of the noise power of the DUT. The cutoff frequency of this equivalent filter is the so-called *noise bandwidth*, Bn.

The digitizer filter is usually a multipole unit with rapid rolloff, and its −3 dB frequency closely approximates Bn. In some tests, though, a simple resistor-capacitor (RC) rolloff may be used. In yet other tests, the device itself has a natural high-frequency rolloff and acts as its own band-limiting filter. If the DUT has a −6 dB/octave (20 dB/decade) rolloff, it has the same noise bandwidth as a single-section RC low-pass filter:

$$B_n = \frac{\pi}{2} \times F_c \ (-3 \text{ dB}) \qquad (8.4)$$

If the −3 dB frequency were 10 kHz, for example, Bn would be 15.7 kHz. A simple software check for digitizer overload is to ensure that none of the A/D converter (ADC) words in the output vector represents an extreme + or code. In the following exercises, assume that the worst-case DUT has an input noise density of 15 nV/root Hz and that the digitizer filter cuts off sharply at 4000 Hz. The combined gain of A1 and A2 = 100,000.

Exercise 3: Find the broadband (4000 Hz) RMS noise voltage from A2.

Exercise 4: Find the largest expected peak voltages from A2. (Hint: Use the relations introduced in Chapter 5.)

Accuracy and Repeatability of Noise Measurements

The signal to be measured is nonperiodic and infinitely continuous. To state its properties with certainty, we would need to analyze it for an infinite time. We will term this a *global* measurement. By contrast, the test engineer is forced to work with limited or localized intervals of this signal (i.e., to make *local* measurements).

Successive local measurements will differ not only from the hypothetical global measurement but also from one another. Every local measurement is thus subject to *intrinsic measurement uncertainty*. The amount of uncertainty is inversely proportional to the square root of the *local interval*.

A DSP system offers another source of uncertainty: It does not see the true, continuous waveform, but only discrete samples. Sampling the global signal gives a global vector, having unlimited N. In some texts, a complete (global) set of discrete elements is termed the *population*.

For the population to represent the continuous global signal accurately, F_s must be greater than 2Bn. This sounds easy, but a DSP system is limited in the length of the vectors it can process. In production testing, one does not have the luxury of sampling both at a high rate *and* for a long time. Yet, reducing either can increase the measurement uncertainty. The confidence we have in a local vector thus rests on two concerns: (1) How well does the local interval represent the true global signal? (2) How accurately do the local vector samples represent the local interval?

In developing a noise test plan, here are some of the questions that should be asked:

1. How trustworthy is the local, sampled measurement as an indicator of the true global, continuous signal?

2. How much uncertainty is permissible in the test plan?

3. How much uncertainty is expected for the chosen test conditions?

4. How many samples should be collected and processed?

5. What is the most effective way to process them?

6. How can one determine if a noise measurement procedure has yielded correct results?

We will answer these questions with the aid of the graphs and formulas that follow.

Statistical Sampling versus DSP Sampling

We must be very cautious when trying to estimate measurement uncertainty by formulas obtained from statistics texts. When statisticians and test engineers use the word ''sample,'' they are talking about two very different things.

Statistics texts deal with *independent samples*. Samples of this kind carry completely unique, nonredundant information. Statistical sampling is *sparse sampling*, which in this tutorial means that there are not enough samples to collect the full information available from the continuous signal. This is analogous to the television viewing polls that interview only a few households. In DSP testing, sparse sampling results from making F_s less than the lowest rate established by the universal sampling rule described in Chapter 3.

DSP, by contrast, typically uses the opposite conditions, since the intent is to capture *all* of the information in the signal presented to the digitizer. Except at the marginal rate, there are more than enough samples to do the job, so they consequently carry overlapping, redundant information. They are *not* statistically independent.

Please do not confuse sparse sampling with *undersampling*, which generally refers to the condition where the signal spectrum is made to reside in an image zone other than band zero (i.e., other than in the Nyquist zone). This makes $F_s/2$ less than the highest frequency of the signal but still greater than the bandwidth. Enough samples are gathered to ensure that the signals are fully characterized.

Noise that is bandlimited from 900 to 1100 Hz can thus be undersampled at a rate less than 2200 s/s. But for informational completeness, F_s must be greater than 400 s/s. To sample the signal without spectral ambiguity, the universal rules require that F_s be greater than 450 s/s.

(In the coherent case of Chapter 4, there appears to be no lower limit on F_s. This is because the spectrum of a completely periodic signal is a group of zero-width lines, and the informational bandwidth is therefore zero. This is not relevant to random noise, of course.)

An interesting aspect of sparse sampling is that only the spectral (frequency) distribution is damaged by the loss of information, not the amplitude distribution. As long as N is fairly large, the amplitude distribution of the vector continues to reflect that of the continuous local signal. *Time-domain parameters and total power can be determined from sparse vectors as well as from conventionally sampled vectors.*

Figure 8.2 shows two local intervals of sparsely sampled, wideband, normally distributed (Gaussian) noise. The digitizer prefilter was removed, so the noise bandwidth greatly exceeds $F_s/2$. This was done to produce vectors with independent elements, permitting analysis by the formulas given in conventional statistics texts.

Estimating the Repeatability of Local Measurements

A fundamental issue in production noise measurement is how well a single local measurement represents the hypothetical global measurement. That is, what is the uncertainty in a single measurement of noise? We need to know this before writing the program to see if it is even worthwhile to attempt the measurement under the proposed test conditions. It also gives us a benchmark in program debugging, to see whether the observed nonrepeatability is reasonable or indicates trouble in the fixture.

To determine the uncertainty with conventional sampling (i.e., where $F_s > 2Bn$), we must first determine the maximum number of statistically independent samples that could have been obtained by sparse sampling.

This number u is derived from the boundary condition between statistical and Nyquist/Shannon sampling, so it also represents the minimum number that could have been obtained by informationally complete sampling if the signal were first heterodyned so that its lowest frequency were 0 Hz. For conventional sampling, u is given by a simple formula relating vector length N, sampling rate F_s, and noise bandwidth Bn. In this instance, Bn is specifically the noise bandwidth that is going to be analyzed, not the broadband seen by the digitizer.

$$u = 2 B_n \times \frac{N}{F_s} \qquad (8.5)$$

Exercise 5: Noise is sampled at 3000 s/s until 1024 samples are collected. The noise band to be analyzed is 800 to 1200 Hz. Find u.

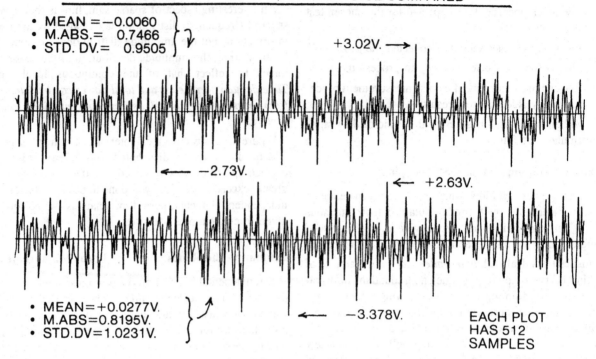

- MEAN = −0.0060
- M.ABS.= 0.7466
- STD. DV.= 0.9505

+3.02V. ⟶

⟵ −2.73V.

⟵ +2.63V.

- MEAN=+0.0277V.
- M.ABS=0.8195V.
- STD.DV=1.0231V.

⟵ −3.378V.

EACH PLOT
HAS 512
SAMPLES

WIDEBAND NOISE SAMPLED AT 10240 SAMPLES/SECOND

Figure 8.2

$$u = 2 * 400 * 1024/3000 \qquad (8.6)$$

$$u = 273$$

Exercise 6: Noise is to be measured in a band from 6 to 14 Hz. Find u if $F_s = 256$ s/s and $N = 512$.

The ratio of the local uncertainty to the local RMS noise depends solely on u. Let S_n represent the standard deviation of the local vector (the RMS noise), and let S_a represent the standard deviation of the *answer* (i.e., the RMS measurement uncertainty). Then,

$$S_a = S_n/\sqrt{2u} \qquad (8.7)$$

Although this is an approximation, it is very close when $u > 30$. The probability that a single local measurement deviates from the average of many local measurements by more than the amount Sa is about 32 percent. The probability that the deviation exceeds 2 S_a is about 4.5 percent; for 3 S_a, it drops to 0.27 percent.

Exercise 7: The measured noise density of a certain op amp is 16 nV per root hertz. The number of independent samples u is 100.

1. Find the RMS uncertainty

2. Find the probable error

3. Find expected range of measurements if the test is repeated 1000 times

Cautions about Averaging

There is a common misconception that an accurate noise measurement is obtained by averaging a large number of local RMS measurements. This probably derives from the fact that *repeatability* is improved by averaging. Given M local measurements, each based on the same effective number of informational samples u, the uncertainty in the average of those measurements is equal to the uncertainty of one measurement divided by the square root of M.

By Equation (8.6), however, we see that the same reduction in uncertainty is accomplished by using a single vector of M * N samples. The reason is that u is increased by the same factor M.

With respect to accuracy, however, averaging is worse than using one long vector. Contrary to intuition, the average does not converge on the correct answer, for two reasons. First, averaging the RMS or p-p measurements instead of the power introduces a fundamental mathematical error because it is done after rather than before the square root operation. The average of 4 W and 16 W is 10 W, for example, and the RMS voltage (relative to 1 ohm) is 3.162 volts. If you started with the RMS voltages, however, you would obtain the average of 2 and 4 V, which is only 3 volts, not 3.162. As was stressed in Chapter 2, in dealing with orthogonal or independent signal components, it is *power* that is additive, not voltage.

Even if one were careful to average power, the noise power obtained by averaging many local measurements still falls short of the true global power. The reason is that removing the individual DC components from each short vector does not produce the same AC sample values as if the DC component of the complete population had been used. The best estimate of the true population power P_t is larger than the average of many local power computations P(average), by the amount,

$$P_t = P \text{ (average) } * \frac{u}{u-1} \qquad (8.8)$$

where u is that of a single vector, *not* the total.

A more precise estimate is obtained if the mean of each local vector is known as well as the AC power. For M local intervals of a given noise waveform, the power of an equivalent single vector of MN time samples is

$$P_t = P(\text{average}) + \text{variance of local means} \qquad (8.9)$$

Of course, this is what you would get directly from one long vector.

Another disadvantage of short vectors is that they do not "see" low frequencies. Remember, the low frequency limit is dictated by the reciprocal of the test period. Using more vectors does not help.

Using a long vector is also better because u is larger, reducing the need to apply the correction of Equation (8.8). If u = 100, for example, the error in using P (average) directly for P_t is only about 1 percent. This translates to an error in the RMS or p-p noise voltage of only 0.04 dB. When only one measurement is made, the computed AC power is itself the best estimate of the average of P(average).

In summary, when precision is required, it is better to work with a long vector than with many short ones. These formulas are provided simply because practical DSP machines are limited in vector processing length, and the desired N may be beyond the machine's single-vector capacity.

Computing Spectral Power from a Sparsely Sampled Signal

Broadband noise power can be estimated closely with sparse sampling. The reason for this is that the amplitude distribution and sample range of samples, sparse or not, are determined by the distribution and range of the continuous signal. For the range of N typically encountered in DSP-based testing, the power of the time series closely approximates that of the continuous signal.

What may be less obvious is that the total power can also be determined from the FFT spectrum of the sparse time vector. The actual FFT distribution is incorrect, of course, because higher frequency components are folded into the Nyquist zone. But folding raises the average noise density, while it reduces the apparent Bn. It has been shown [Ref. 7, pp. 244-245] that the power appearing in this compressed band is equal to that of the original broadband signal.

The relationship is proven only for an infinite sampling interval, but with typical N there is excellent agreement. The effective or apparent noise bandwidth Bn is $F_s/2$. By Equation (8.5), this is consistent with the intuitive relation that sparse N = u. In Figure 8.2, F_s is 10240 s/s and the effective Bn is 5120 Hz.

Consider a test plan that calls for the total noise power over the range, 0.1 to 1000 Hz. Shannon sampling calls for a sampling rate over 2000 s/s. Over the required minimum 10 second test interval, this would produce over 20,000 samples.

The measurement can nonetheless be made successfully with just a few hundred samples. F_s might be 102.4 s/s, for example, so that N = 1024.

Now suppose another test calls for noise density measurement at 0.2 and 1 Hz. This can also be done at the above rate, but density ideally requires Shannon (complete) sampling. For this, the digitizer filter would have to block frequencies over 50 Hz.

Can the two tests be combined? Opening the digitizer filter range to admit components up to 1000 Hz will cause folding or aliasing. But in many semiconductor amplifiers, the low

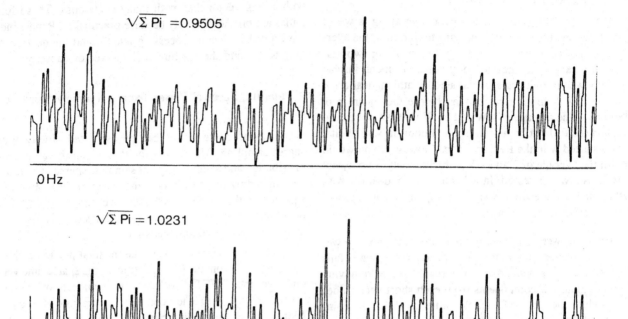

$\sqrt{\Sigma \, Pi} = 0.9505$

$\sqrt{\Sigma \, Pi} = 1.0231$

.0 Hz

Figure 8.3

frequency noise density is so much larger than the aliased components, that the increase in low frequency density may be negligible. In such cases, sparse sampling is still acceptable.

The key in such methods is to make sure that $F_s/2$ extends into the high frequency region where the relative noise density is small when compared with that in the bands of interest. For 0.1 and 1 Hz noise, the above cutoff frequency of 50 Hz might well be sufficient.

Exercises

Several plots of noise signals and spectra are shown in Figures 8.3-8.11 . Use the preceding relations to see if these signals are actually behaving according to theory, in terms of power, voltage, bandwidth, and repeatability.

Figure 8.3 shows the spectral plots of the two local time intervals shown in Figure 8.2. The horizontal axis is fre-

quency, from 0 to 5120 Hz. Amplitude (voltage density) is plotted vertically.

These spectra were obtained by using the magnitude FFT without special windowing. Even though the signals are not periodic, windows are generally not needed when the FFT is applied to Gaussian or near-Gaussian vectors of typical DSP lengths.

Such techniques are most often required when the signal is strongly but imperfectly periodic, as when a periodic signal is corrupted by another periodic signal that is not coherent or when its period is not an exact submultiple of the FFT period. An example of a windowing technique suitable in the latter case is discussed by Rosenfeld in the article, "DSP Measurement of Frequency," in Chapter 13.

The spectral distribution in Figure 8.3 is not trustworthy, of course, because the time vector was sparsely sampled. If it were required to find the noise density in specific bands, then the incoming noise would have to be band limited to

AMPLITUDE RESPONSE OF BANDPASS FILTER
USED IN NOISE MEASUREMENT EXPERIMENT

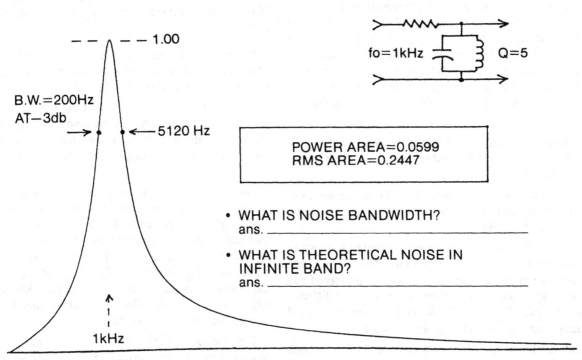

B.W.=200Hz
AT−3db

POWER AREA=0.0599
RMS AREA=0.2447

- WHAT IS NOISE BANDWIDTH?
  ans. _____
- WHAT IS THEORETICAL NOISE IN
  INFINITE BAND?
  ans. _____

Figure 8.4

TWO INTERVALS OF NARROW-BAND NOISE

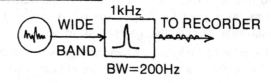

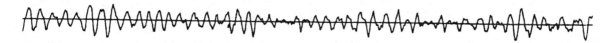

- MEAN=0.0000
- RMS=0.2096
- RANGE=−0.5384 TO +0.5673

- MEAN=0.0001
- RMS=0.2737
- RANGE=0.7768 TO +0.7129

←———————————— 50ms ————————————→

Figure 8.5

less than $F_s/2$. In the present example, though, we are comparing only the power and the uncertainty.

Figure 8.4 shows the response of a computer model of a resonant, bandpass filter with a Q factor of 5. This indicates that its center frequency is 5 times the half-power bandwidth. The computed power area and RMS area of the plotted response curve are shown on the graph, and apply to the range, 0 to 5120 Hz, which is the Nyquist band for the previous noise. These values are relative to a brick wall filter with fc = 5120 Hz and therefore indicate the amount by which this filter will attenuate the power and voltage of broadband 5120 Hz noise.

Exercise 8: Compute the noise bandwidth of this filter in Hz. You may use the RC rolloff formula by substituting bandwidth for Fc.

Exercise 9: Calculate the ratio, Bn/5120 Hz. Compare this with the computer-calculated power area shown on the graph. Explain the difference.

In Figure 8.5, this filter curve was applied in software to each of the two previous noise vectors. This was done to ensure that all of the changes between the unfiltered and filtered signals were due to filtering, not to using different noise intervals. The mean, RMS, and range were calculated directly from the filtered time series vector.

Exercise 10: How do the two narrow-band RMS voltages compare with the values predicted by the RMS area of the filter?

Exercise 11: Note that the variation between the two local RMS values is much larger in ratio or percentage for the two filtered waveforms than for the two unfiltered waveforms. As a general principle, why is this so? What is the quantitative formula that governs this effect?

Exercise 12: Why are the mean voltages so much smaller in the filtered waveforms than in the unfiltered ones?

Figure 8.6 shows the spectra of the two filtered time vectors, derived by magnitude FFT. In each plot, the bin power was summed over the Nyquist band, and the square root applied to find the RMS voltage.

Compare these two calculated values with the RMS voltages computed from the time series vector. The closeness of results again suggests that, for power measurement of random noise, the FFT does not require windowing or other special estimation techniques.

Figure 8.7 shows the plot of a local interval of noise that has been filtered by a narrow bandpass 1 kHz filter (Q=50). Note that the output resembles an amplitude-modulated 1 kHz carrier.

Exercise 13: Explain the AM-like behavior by the effect of the filter on the signal spectrum. (Hint: Consider the spectral lines adjacent to the 1 kHz line.)

Exercise 14: Suppose the waveform were clipped close to the horizontal axis so only the zero crossings were visible. Use the above sideband relationship to demonstrate that the crossings are not completely regular in spacing.

The final charts in this chapter (Figures 8.8-8.11) are four statistical print-outs of noise density measurements of an LF355 op amp, supplied by Kostas Tsakmaklis of LTX. In each bar graph, the "sample size" indicates the number of measurements made on the DUT with the specified bandwidth, sampling rate, and N. The length of each bar shows the relative frequency of occurrence of the RMS noise density shown at the right. The actual number and the cumulative percent are shown at the left.

The heading, "standard deviation," is the parameter previously denoted S_a, in Equation (8.7); this parameter is the observed RMS uncertainty of the answer. The true or population RMS is not shown but, as you may recall, is slightly higher than the mean of the local RMS readings. The best estimator of the population standard deviation, when only the mean of the local standard deviations is known, is

$$S_p = S_n \text{ (average) } \sqrt{\frac{2u}{2u-3}} \qquad (8.10)$$

Most of these statistical relations depend on what is called the q-distribution, given in the *Handbook of Probability and Statistics with Tables*, Second Edition, by Burington and May, 1970, McGraw-Hill. As in most statistics texts, the number of samples n is not the number of DSP samples, but the number of independent (informational) samples denoted in this tutorial by the letter u.

SQUARE ROOT OF POWER=0.2096

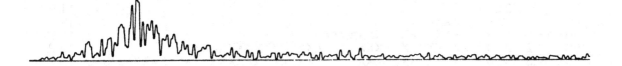

SQUARE ROOT OF POWER=0.2737

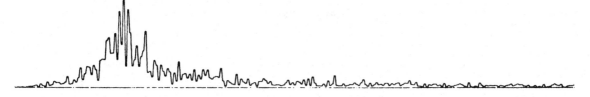

Figure 8.6

NOISE FILTERED BY NARROW BANDPASS FILTER

fc=1KHz, Q=50

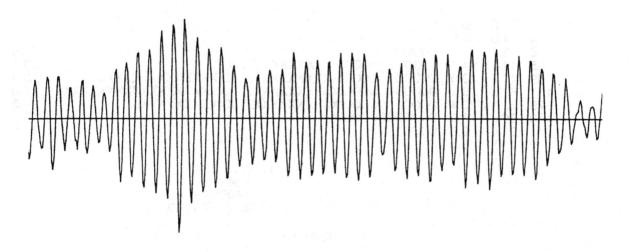

RMS VOLTAGE=0.0862

Figure 8.7

STATISTICAL PLOT

PROGRAM: 3551024S.KT REVISION: 12:00 PM 10/22/84

TEST 200.5 SPECTRAL NOISE AT 1000 Hz SAMPLE SIZE 263

MEAN 13.680 NV STD DEV 0.711912 MIN 12.214 MAX 16.280

```
TOTAL CUM%    %                                                  ALL VALUES IN NV

  0   0.0   0.0  I                                                 <   11.9
 10   3.8   3.8  IXXXXX>                                               12.2
 30  15.2  11.4  IXXXXXXXXXXXXXXXXXX                                   12.7
 58  37.2  22.0  IXXXXXXXXXXXXXXXXXXXXXXXXXXXXXXXXXXX                  13.2
 79  67.3  30.0  IXXXXXXXXXXXXXXXXXXXXXXXXXXXXXXXXXXXXXXXXXXXXXXXX     13.7
 49  85.9  18.6  IXXXXXXXXXXXXXXXXXXXXXXXXXXXXX>                       14.2
 29  96.9  11.0  IXXXXXXXXXXXXXXXXX>                                   14.7
  4  98.4   1.5  IXX                                                   15.2
  1  98.8   0.3  I>                                                    15.7
  3  100.   1.1  IX>                                               >   15.9
                                  DEVICE #24
                                  N=1024
                                  FS=3000 S/S
                                  W=900 TO 1100 Hz
```

Figure 8.8

STATISTICAL PLOT

PROGRAM: 355ALL9.KT REVISION: 12:01 PM 10/19/84

TEST 200.5 SPECTRAL NOISE AT 1000 Hz SAMPLE SIZE 386

MEAN 13.8940 NV STD DEV 0.728265 MIN 12.120 MAX 16.152

```
TOTAL CUM%    %                                                  ALL VALUES IN NV

  0   0.0   0.0  I                                                 <   11.7
  4   1.0   1.0  IX>                                                   12.0
 16   5.1   4.1  IXXXXXXX                                              12.5
 55  19.4  14.2  IXXXXXXXXXXXXXXXXXXXXXXXXX>                           13.0
 89  42.4  23.0  IXXXXXXXXXXXXXXXXXXXXXXXXXXXXXXXXXXXXXXXXX            13.5
101  68.6  26.1  IXXXXXXXXXXXXXXXXXXXXXXXXXXXXXXXXXXXXXXXXXXXXXXX      14.0
 79  89.1  20.4  IXXXXXXXXXXXXXXXXXXXXXXXXXXXXXXXXXXXXX                14.5
 29  96.6   7.5  IXXXXXXXXXXXXX                                        15.0
 11  99.4   2.8  IXXXX>                                                15.5
  2  100.   0.5  IX>                                               >   15.7

                                  DEVICE #24
                                  N=2048
                                  FS=4096 S/S
                                  W=900 TO 1100 Hz
```

Figure 8.9

STATISTICAL PLOT

PROGRAM: 3551024S.KT   REVISION: 12:00 PM   10/22/84

TEST 300.5   SPECTRAL NOISE AT 100 HZ   SAMPLE SIZE 263

MEAN 20.233   NV   STD DEV 2.576579   MIN 12.227   MAX 28.775

TOTAL CUM%   %   ALL VALUES IN NV

```
 1    0.3   0.3   IX                                              ‹    12.5
 0    0.3   0.0   I                                                    13.0
 3    1.5   1.1   IXXX                                                 14.0
 5    3.4   1.9   IXXXXX                                               15.0
11    7.6   4.1   IXXXXXXXXXX                                          16.0
14   12.9   5.3   IXXXXXXXXXXXX                                        17.0
31   24.7  11.7   IXXXXXXXXXXXXXXXXXXXXXXXXXXXX                        18.0
45   41.8  17.1   IXXXXXXXXXXXXXXXXXXXXXXXXXXXXXXXXXXXXXXXXXXX  19.0
37   55.8  14.0   IXXXXXXXXXXXXXXXXXXXXXXXXXXXXXXXXXX                  20.0
36   69.5  13.6   IXXXXXXXXXXXXXXXXXXXXXXXXXXXXXXXXX                   21.0
31   81.3  11.7   IXXXXXXXXXXXXXXXXXXXXXXXXXXXX                        22.0
21   89.3   7.9   IXXXXXXXXXXXXXXXXXXX                                 23.0
12   93.9   4.5   IXXXXXXXXXXX                                         24.0
10   97.7   3.8   IXXXXXXXXX                                           25.0
 4   99.2   1.5   IXXXX                                                26.0
 1   99.6   0.3   IX                                                   27.0
 0   99.6   0.0   I                                                    28.0
 1  100.    0.3   IX                                              ›    28.5
```

                        DEVICE #24
                        N=1024
                        FS=3000 S/S
                        W=60 TO 140 Hz

Figure 8.10

STATISTICAL PLOT

PROGRAM: 355ALL9.KT   REVISION: 12:01 PM   10/19/84

TEST 300.0   SPECTRAL NOISE AT 100 HZ   SAMPLE SIZE 386

MEAN 19.914   NV   STD DEV 1.736889   MIN 15.457   MAX 26.993

TOTAL CUM%   %   ALL VALUES IN NV

```
 0    0.0   0.0   I                                              ‹    14.5
 1    0.2   0.2   I›                                                  15.0
 9    2.5   2.3   IXXXX›                                              16.0
14    6.2   3.6   IXXXXXXX                                            17.0
58   21.2  15.0   IXXXXXXXXXXXXXXXXXXXXXXXXXXXXXX                     18.0
86   43.5  22.2   IXXXXXXXXXXXXXXXXXXXXXXXXXXXXXXXXXXXXXXXXXXXX  19.0
86   65.8  22.2   IXXXXXXXXXXXXXXXXXXXXXXXXXXXXXXXXXXXXXXXXXXXX  20.0
60   81.3  15.5   IXXXXXXXXXXXXXXXXXXXXXXXXXXXXXX                     21.0
41   91.9  10.6   IXXXXXXXXXXXXXXXXXXXX                               22.0
25   98.4   6.4   IXXXXXXXXXXXX                                       23.0
 4   99.4   1.0   IXX                                                 24.0
 2  100.    0.5   IX                                              ›    24.5
```

                        DEVICE #24
                        N=2048
                        FS=4096 S/S
                        W=60 TO 140 Hz

Figure 8.11

Chapter 9
Introduction to A/D Testing

Chapter 9: Introduction to A/D Testing

Chapter 10 of this tutorial covers digital signal processing (DSP) techniques for testing high-speed analog-to-digital converters (ADCs), and is intended for the test engineer experienced in using and/or testing such devices. This chapter supplies preliminary information for the reader who is new to this area, or whose only exposure to ADC testing has been through textbooks and technical articles. Additional support information is provided by the references listed in the appendix, under the title, "Reference Articles for DSP and A/D/A Testing."

A/D versus D/A Conversion

A key consideration in converter testing is that an ADC is not the mathematical inverse of a digital-to-analog converter (DAC). What this means is that DAC testing requires a different set of test procedures than ADC testing and that parameters such as integral linearity and differential linearity are defined differently for the two types. The equations that apply to one type cannot simply be turned around (i.e, solved in reverse) to provide equations for the other type, as is possible with complementary linear transfer functions.

Popular articles often fail to point out this distinction. Unfortunately, there are just enough similarities between the two converter types that the same techniques and same definitions may initially appear to work in both cases, but the result of such imprecise thinking is often that an entire production run is judged incorrectly. The fact is that ADC techniques such as histograms, MIL-SPEC regression methods, and cosine regression do not *directly* apply to DAC. Likewise, D/A methods like Walsh-Rademacher techniques and input-output correlation methods require some redefinition to be used in ADC testing. There are analogous methods on opposite sides of the fence, to be sure, but the relation is not as simple as it might first appear.

Consider a transmission channel composed of an ideal sampler (track-and-hold circuit, or sample-and-hold circuit) and an ideal reconstruction filter (Figure 9.1). Chapter 3 showed how proper reconstruction filtering removes the errors of stepwise signals and produces a perfect, continuous replica, assuming the steps are correct in amplitude. As long as the rules of sampling are obeyed, these two blocks behave as inverse linear transfer functions: The channel will not distort the signal. No information is lost.

The trouble begins when an ADC and a DAC are added between the sampler and the filter, creating a digital link (Figure 9.2). Distortion is introduced into the output waveform, *even if the functions are ideal.* (Texts often speak of quantization "noise" but this distortion is not like the random noise discussed in Chapter 8.)

Since the ADC and DAC are ideal, distorted transmission implies either that the ADC and DAC are not inverse functions or that the process is intrinsically incapable of transmitting the total information of the input signal. In fact, both of these implications are true. An ideal ADC discards information, while an ideal DAC does not. This fact alone tells us that at the statistical level, A/D and D/A test procedures and parameters should differ in some fundamental way.

Transfer Maps

Certain transformations can be *mapped* on a one-to-one basis: Given a discrete set of input states, there exists one and only one corresponding output state. Such one-for-one mapping can be shown on a *point map*, where each point connects the input state with its unique response state. Ideal DAC transfer is described by a point map (Figure 9.3).

The ideal DAC has no transfer ambiguity. If we are told the states only at one end—either end—we know what the

Figure 9.1: Distortion-free channel

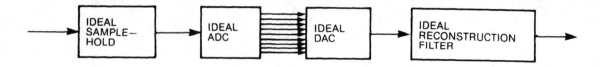

Figure 9.2: Distortion-producing channel

states at the other end must be. Technical articles frequently use staircase waves as DAC transfer curves, but the ideal map has only points.

A/D conversion, on the other hand, cannot be described by a point map. Like the ideal DAC, the ideal ADC has a discrete output set. But for each code out, there is a fuzzy input, a *continuum* of input voltage. The ADC has *one-way uncertainty*. If a specified analog level is applied to the input of a perfect ADC, we know precisely what its output code response will be. But, if we are told, instead, only what the output code state is, we cannot tell the exact input voltage, only its range. The uncertainty is uniformly distributed over the width of the step, or the least significant bit (LSB), or the quantum size, whichever you prefer (Figure 9.4).

It is a tenet of ADC testing that one does not test an ADC by applying DC voltages (or steps) and looking to see if the ADC responds with the right or wrong codes! The exclamation point is there because it is likely that, even as you read this, there are large numbers of ADCs being tested this way. Nonetheless, it is a rather poor way of doing things. Figure 9.5 shows a case in which the same nominal input voltages are applied to two different ADCs, where both respond with the "correct" codes. As you can see, however, the two units are quite different in the actual shape of their transfer curve.

A more significant hazard of treating an ADC as though it had a point map, is that real ADCs will usually appear to give many wrong codes, even though the units may actually be within specifications. Suppose, for example, the gain of the input buffer is slightly higher than normal. This in itself might be perfectly acceptable, since it might be intended that the user adjust the ADC reference voltage in the actual application. But if this unit is tested by simply applying steps of voltage and comparing the resulting code words against the ideal, the additional gain will cause the slope of the transfer curve to increase; voltages applied at the right side of

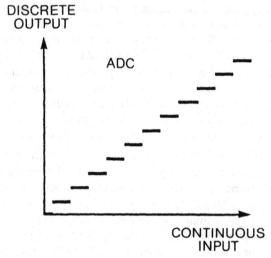

Figure 9.4

the graph will produce code levels higher than the nominal levels, making the ADC appear to be bad.

Voltage-to-code comparison is also affected by input offset. Untrimmed DC offset increases the amount of apparent error, causing *all* code steps to shift positions.

The problem is compounded when sampling converters are dynamically tested, where the input steps are those of a sampled sinusoid. Even if the samples follow the waveform faithfully, any phase shift in the ADC analog path (probably not specified at all) will cause the samples to fall at the different points on the sinusoid than the expected points, making a

INPUT-OUTPUT MAP OF
IDEALIZED DIGITAL-TO-ANALOG CONVERTER

DISCRETE ANALOG OUTPUT LEVEL

DIGITAL INPUT CODE LEVEL

Figure 9.3

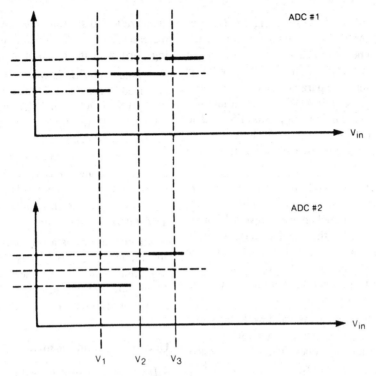

Figure 9.5

one-for-one comparison of output code words with the expected words meaningless.

Another problem is that whether the testing is static or dynamic, the steps shown in the previous figures as solid bars with definite edges are in reality fuzzy lines. Because of random noise, these code edges, or *decision levels*, are only probabalistic locations.

Repeating a test most likely will give a different set of codes, especially in high-resolution converters, where the quantum size is not much greater than Boltzmann's noise. The device under test (DUT) that appears to pass one time might appear to fail the next, or vice versa.

It is clear that we can identify and correct for all of these contributing factors—gain, offset, phase, and noise—by looking at the full collection of ADC codes in context (i.e., by analyzing the set as a *vector* not as isolated code words). *ADC testing thus involves more statistical work than DAC testing.* The wrong way is to try to do what is done in digital testing, where the output code words are compared with an expected code set, one by one.

In summary, because the DAC has a point map, DAC linearity tests can take advantage of input-output relationships that are fairly simple. Because the ADC map has fuzzy bars, however, ADC testing requires a lot more samples, and generally more complex analysis, than a DAC of the same resolution.

Transmission Parameters versus Intrinsic Parameters

A *parameter* is a numerical value that characterizes a component or a system. Modern ADC and DAC testing looks at two types of parameters: transmission and intrinsic.

Transmission parameters characterize the channel in which the ADC or DAC under test is embedded, usually by its affect on a conventional (sinusoid or multitone) test signal. Examples include gain, signal-to-distortion, intermodulation (IM) distortion, noise power ratio (NPR), differential phase shift, and envelope delay distortion. In CCITT standards, transmission parameters are known as *performance parameters*.

In non-DSP testing, those portions of the channel other than the ADC or DAC under test are formed from laboratory-quality "reference" hardware, so the input and output signals are real analog waveforms. Analysis is performed by conventional instruments distortion analyzers.

In DSP testing, by contrast, the reference circuits are usually software models. To test a DAC, the digital input pattern is prepared in advance, emulating the output of an ideal sampling ADC with an ideal analog test input. To test an ADC, the appropriate analog input signal is supplied by the system digitizer and a response vector is collected from the ADC digital output by the system receive memory. In essence, the ADC under test acts as its own digitizer.

Intrinsic parameters characterize the DUT itself, indicating some ideally invariant circuit property: something that exists internally, independent of the input signal. The full scale range (FSR) of ADC input voltage is one such parameter; each unit can be said to have a certain FSR even though no signal is applied at the moment. In truly linear circuits, *gain* is another intrinsic parameter; it is a ratio that does not vary with signal size. For ADCs, the most familiar intrinsic parameters are the number of bits, the FSR, static linearity (differential and integral), maximum clock rate, and the code format. For DACs, the list includes settling time and glitch area.

Years ago, virtually all ADC and DAC specifications were phrased in terms of intrinsic parameters. With low production volume, the manufacturer could make "universal" units, without knowing—or limiting—the end applications. From intrinsic specifications, the user could estimate how the device would perform in actual use. It also meant that the test method was not critical, since the parameters (in theory) did not depend on test conditions.

Today's converter usage is so great that new designs tend to be applications-specific and transmission parameters are often more helpful than intrinsic ones. This is particularly true as frequencies and conversion rates move ever higher, and parameters such as linearity *are no longer invariant*. This latter fact also implies that the test method is increasingly critical.

Conversion Formats and Types

Most of the techniques in this tutorial apply to so-called *linear converters*. Strictly speaking, this is a contradiction in terms, since A/D conversion is an intrinsically nonlinear transfer process. What we really mean when we say "linear" ADC is that there is no nonlinearity other than that of quantization. All steps are equal in width. A curve passed through the *code centers*—the statistical centers of each fuzzy bar— would be a straight line.

Mu-law and A-law encoders used in digital telephone systems are (deliberately) nonlinear ADCs. They provide linear piecewise approximations to a logarithmic transfer curve.

Floating-point (FP) converters are also deliberately nonlinear. The code word is divided into two distinct groups, plus a sign bit. The low-order group is generated by a linear ADC. The high order represents an exponent, usually (but not necessarily) of the base 2. The base, raised to the exponent, indicates the analog amplification necessary at the matching DAC to linearize the overall A/D-D/A transfer function. A mu-law ADC is a FP converter with 4 linear bits, 3 exponent, and 1 sign bit. The decoding DAC has 8 binary gain ranges.

The *dynamic range* of an FP converter is much larger than the *signal-to-distortion* (S/D) *ratio,* because the latter is governed mostly by the mantissa bits. The absolute level of quantization distortion power increases as the exponent increases, so the S/D ratio does not vary much over the working portion of the dynamic range.

A *binary-coded decimal* (BCD) converter has its bits arranged in groups of 4, each group representing a decimal digit, 0 to 9. A 16-bit BCD converter thus uses only 10 * 10 * 10 * 10, or 10000 code states. It uses only a fraction of the unique code states.

In this tutorial, a *binary converter* means one that uses all binary code combinations. A 16-bit binary-coded converter, unless otherwise indicated, has 65,536 unique code states. Its resolution is thus more than six times better than that of a 16-bit BCD converter.

If a converter has a *restricted* range of codes, the effective number of bits (so far as resolution is concerned) is the logarithm to the base 2 of the number of code levels. A 16-bit BCD converter thus has the resolution of a 13.29-bit binary converter. The binary logarithm equals the common log times 3.322.

Uncertainty and Distortion of the Ideal ADC

If we rotate Figure 9.4 so that the transfer slope is horizontal, we shall see that the error function is a sawtooth wave. The RMS *uncertainty* of the quantization, seen by an observer looking back into the output of the ADC, is the RMS amplitude of this sawtooth. This amplitude is also the RMS value of the *distortion* introduced into an analog signal that is transmitted through the channel of Figure 9.2. The p-p height of the sawtooth is 1 LSB, and the power of this waveform is 1/12 LSB squared. If Q represents the quantum (LSB) voltage, then the RMS quantization distortion voltage is

$$D = \frac{Q}{\sqrt{12}} \text{ volts, RMS} \qquad (9.1)$$

In an n-bit linear binary ADC, there are 2^n code levels in the FSR. The two end steps have no outer bounds, of course, and so do not actually have statistical centers. In tests that call for real centers (e.g., the MIL-STD flash converter tests), these two end codes are discarded.

To analyze ideal behavior, however, we may assume centers for all code steps if we adopt the convention of AT&T and assign *virtual edges* (i.e., hypothetical decision levels placed where edges would exist if the transfer curve were to continue). These virtual edges are the *clipping levels* of the ADC. Let us accordingly apply a hypothetical sinusoid to the ideal ADC input such that its peaks just touch these virtual edges.

The RMS amplitude is, then,

$$\text{FS Sine Amplitude} = \frac{Q \times 2^n}{\sqrt{8}} \text{volts, RMS} \qquad (9.2)$$

Dividing this into the RMS quantization distortion gives the ratio,

$$\left[2^n \times \sqrt{1.5}\right]^{-1} \qquad (9.3)$$

and, expressing this in dB, we obtain

$$\text{Relative Distortion Level in dB} = -(6.02n + 1.761) \qquad (9.4)$$

This is *not* the signal-to-noise ratio of the ADC, but it is the amount of ADC distortion *relative to the level of a hypothetical full-scale sinusoid*. The formula applies exclusively to linear conversion.

In linear ADCs, the distortion power is essentially independent of the signal level so long as quantization is actively taking place (i.e., so long as a signal is applied that is at least several LSB in amplitude). It can therefore serve to indicate the approximate signal-to-noise ratio, if it is adjusted to reflect the actual power level of the test signal, which may not be a sinusoid or may not be at full scale.

Note also that the computed distortion level is not quite the same thing as the *measured level*, because some portion of the distortion power will fall into the same fast Fourier transform (FFT) bin (or bins) as the test tone (or tones). Only that portion in otherwise empty bins can be discerned as distortion or "noise" in the typical DSP test. This division of quantization power was illustrated in Chapter 4 (Plates 4.9 and 4.10), by using a 6-bit ADC with "bad" and "good" M and N values. The problem is even more pronounced with hardware methods because they are less selective than DSP.

Exercise 1: Find a ratio of sampling rate to test tone frequency that will cause the *measured* S/D of an ideal ADC to be infinite.

Exercise 2: Find the relative distortion level in dB when a full-scale sawtooth wave is used as the reference.

DAC Transfer Error

Most readers are familiar with the various kinds of DAC error, but it may prove helpful to relate them graphically to the point map of an ideal DAC. On the point map, all errors must appear as *vertical displacements* from the ideal locations.

Figure 9.6 shows the graphical result of *offset*, a uniform displacement of all points. In the past, this was often measured by applying the code for zero volts and measuring the DC output. But what if that code is out of place, and no other? Or what if the DAC is tested dynamically with signals that do not exercise the zero code? In such cases, offset is more meaningfully defined by the *vertical intercept* of the best-fit straight line with the Y axis.

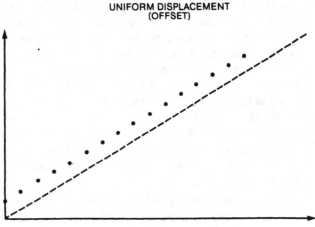

Figure 9.6: Uniform displacement

Figure 9.7 shows another so-called linear error: improper gain, or slope, or FSR. In this case, the vertical displacement is proportional to the input code. With a real DAC, where the points do not precisely follow a straight line, the *gain is the slope of the best-fit line*. The past method of using the two end points (only) to determine the FSR should not be trusted for accurate slope measurement, especially in dynamic testing. The slope of the best fit line is a better indicator. Fast vector mathematics for linear regression line determination is given in Chapter 10.

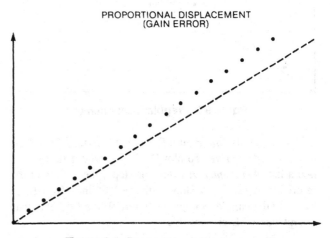

Figure 9.7: Proportional displacement

DAC nonlinearity has two significant aspects: differential linearity error (DLE) and superposition error. *Differential error* (Figure 9.8) is a measure of the irregularity or nonuniformity in the increments of output voltage or current. In the example unit, the irregularity is most obvious at the center of the transfer curve, the most significant bit (MSB) transition. This is where the MSB is switched on and the lower-order bits are switched off. In this unit, the MSB contribution is too small, so the output drops instead of rising at this transition.

The *DLE function* is a step-by-step list of the differences between the actual increments and the linearized increments (i.e., the vertical projections onto the best-fit reference line). Thus, *every* step in the map of Figure 9.8 has differential error. Every step is larger than its linearized value, except for the MSB transition step. That is why the slope of each half is greater than that of the reference line. But the *sum* of all the differential errors is zero in this unit, and there is no overall gain or FSR error. This is always true in DACs in which the bit-weight ratios are constant, even though incorrect. Then, and only then, does the best-fit reference line pass through the end points.

In the example of Figure 9.8, the transfer curve is also *nonmonotonic*. That is, it does not have the same *sign* of its derivative over its full range. Observe that a DAC can have very poor linearity and yet be monotonic as long as the changes are always in the same direction.

IRREGULAR DISPLACEMENT
(DIFFERENTIAL NON-LINEARITY)

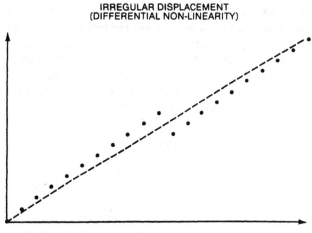

Figure 9.8: Irregular displacement

Note carefully that Figure 9.8 is not a plot of DLE, but of the transfer curve. To plot DLE, you would first calculate a list of *incremental values* or steps, then subtract the reference step size—the slope of the best-fit line—from these steps. In the simple example of Figure 9.8, all differentials except the MSB are positive.

Sometimes the reference step size is called the "nominal" step. Correctly speaking, however, "nominal" refers to the design-center value for all units of one type. Individual units will have higher or lower slope than the nominal, so the reference must be calculated anew for every unit. For a unit in which the bit-weight error is constant, the reference slope (reference step size) is the *average* of the measured step sizes for that unit.

Exercise 3: Sketch the DLE for the simple DAC of Figure 9.8, assuming constant bit-weight error. As a check on the correctness of your sketch, note that the *area* under the sketched curve must be zero. (Why?)

Superposition Error

Superposition error is the result of *nonconstant bit weights*. It is seldom listed separately, but if the test is honest and thorough, shows up as part of the integral linearity error (ILE). Figure 9.9 shows the effect of superposition error on the ILE of an otherwise ideal DAC.

Superposition, by itself, refers to the fact that the current or voltage of a DAC is a linear summation of many components that are ideally independent of one another. Switching the MSB off and on, for example, should not affect the amplitude of the LSB current or any other component. Complete independence is impossible to achieve in real designs, however, so the increment of DAC output depend to some extent on what other bits are activated. Figure 9.9 shows a DAC in which the high order bits "steal" current from the LSB (or perhaps decrease the gain of the output amplifier or buffer) so that the LSB becomes smaller as the code level increases.

ILE is not shown in these plots, but is simply the difference between the actual transfer curve and the reference line. It can be visualized by rotating Figure 9.9 clockwise so that the reference line is horizontal. To produce such a plot, you must first determine the best-fit line (methods are given in Chapter 10) and then project the code levels onto this line to determine the absolute linearized output voltages of the particular DAC under test. These are subtracted from the measured output values, normalized relative to the LSB size, and plotted as a function of code level.

Integral error is aptly named because each point in the ILE vector can be obtained by integrating the differential error vector from the lowest code up to code in question. The first ILE point is the constant of integration, which is assumed to be zero. The next point equals the first DLE value. The next point is the sum of the first two DLE values. The next

DEPENDENT DISPLACEMENT
(SUPERPOSITION ERROR)

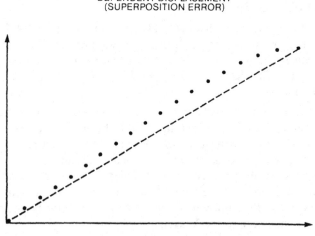

Figure 9.9: Dependent displacement

is the sum of the first three DLE values, and so on. Since the average DLE is zero by definition, the last ILE point plotted this way falls at the same level as the first. (Question: does this mean the reference line must pass through the end points?)

An ILE curve obtained by progressive integration is never identical to one obtained directly from the transfer curve of DLE. For one thing, we have assumed an arbitrary constant of integration; for another, the summations involve many imperfectly measured increments. But the results are very similar, and you will see both used in production testing. It is helpful to try both methods as a check on the correctness of the program instructions.

If a DAC has no superposition error, it is possible to measure the DLE and ILE just by measuring the differential errors at the *major transitions*. A 12-bit DAC, for example, has only 11 such transitions. One method is shown in my article, "Automated Measurement of 14-to-16-Bit Converters," in Chapter 13. Be careful, however, because many modern designs have considerable superposition error, especially segmented designs. Designs that "guarantee monotonicity" are often those that are prone to curve bending caused by superposition error.

I have measured a number of such units that claimed very good ILE, yet which, when measured for *all* codes, exhibited poor ILE because of superposition error. This was very likely because of the use of major-transition-only testing, rather than all-codes testing.

The final form of error considered in this introduction is time-related error (Figure 9.10). The actual voltage level is never steady but is, first of all, subject to random (Boltzmann's) noise. The transfer map of a real DAC does not have points but fuzzy vertical lines. Down at the microvolt level,

if we were to sample the so-called static output repeatedly, the sample set would show a roughly Gaussian amplitude distribution (Figure 9.10).

Some designs, such as certain ones based on capacitor networks, have *hysteresis*. If the point is approached from a lower code, the mean of the little Gaussian curve might be slightly lower than if the previous code word had been a higher code (Figure 9.10).

If the DAC map is considered under dynamic conditions, there is also a transient vertical change when a new code level is first selected. First, there is usually a "glitch," or spike that results from inexact timing between currents that are switching off and those that are switching on. In stepping from code 0111 to 1000, for example, if the high-order bit is faster than the other three, there will be transient state of 1111. If the DAC has a very high bandwidth, the glitch amplitude would be 3 LSB. The transient state change is $3 \rightarrow 7 \rightarrow 4$. Glitches can also be the result of digital-to-analog crosstalk.

Second, the glitch may institute *ringing* or momentary oscillation if there is resonance or if there is a transmission line termination error. Third, the natural RC rolloff of the various circuits introduces a delay in the *settling* of the output current or voltage to the steady state. Finally, there may be a small, but much longer settling component because of thermal time constants in the resistors.

Figure 9.11 illustrates the combined effects, as might be obtained by a large number of samples taken at various times within each step. A good DSP test technique will allow us to start with this composite behavior and to work backwards analytically to isolate the individual components. One method, using Rademacher functions, is shown by Souders and Flach in their article, "An NBS Calibration Service for A/D and D/A Coverters," reprinted in Chapter 13.

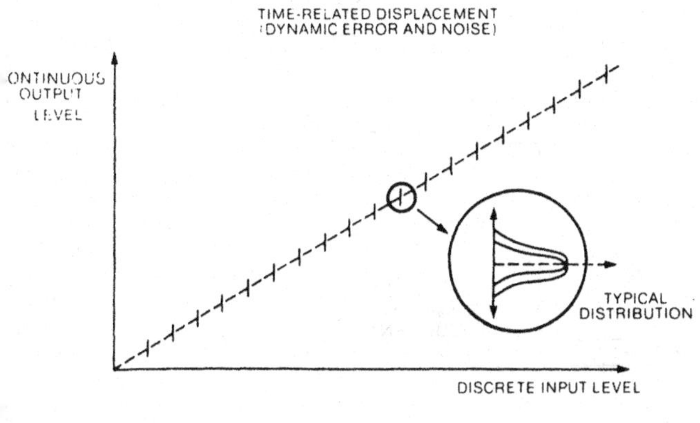

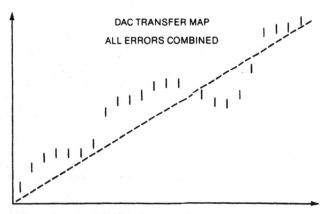

Figure 9.10: Time-related displacement

Figure 9.11: DAC transfer map

Adapting D/A Parameters to ADC Measurement

The voltage levels produced by an ideal DAC are analogous to *code centers* in an ideal ADC. They differ, of course, in that DAC output levels are physically measurable, while ADC code centers are statistical concepts. But the analogy is useful in applying the previous graphical concepts to ADC parameters. ADC ILE, for example, is the plot or function that shows the code-by-code deviation of the step *centers* from the best-fit straight line (Ref. [18], A/D/A references in Chapter 14).

There is another set of ADC voltages, however, that have no corresponding or analogous values in the DAC map. These are the *decision levels* or *thresholds*. From a hardware viewpoint, we could say (generally) that ADCs have comparators; DACs do not.

As a result, differential linearity is defined differently for ADCs. If the decision levels are known, we compute the statistical step (quantum) size for every code and then subtract the average step size to obtain the list or plot of DLE values. By this definition, the DLE cannot be less than -1, in which case the code is said to be *missing*. Contrast this with the DLE definition given for DACs.

A common method for finding ADC DLE without actually finding the code edges is the code *histogram*; that is, a *tally*, or count, of how many times each digital output code appears in the response vector. If the analog input is a linear ramp that covers more than the FSR, the tally for each code will be proportional to the statistical step width. Subtracting the average tally directly produces the DLE vector.

Such procedures are discussed more completely in Chapter 10. To complete this present chapter, let us conclude with a brief look at a traditional (non-DSP) approach to ADC testing, based on the use of a servo loop (Figure 9.12).

The servo loop technique is a hardware method for locating the statistical code edges. The reference digital code is placed into register B and the nominal DC voltage corresponding to the code center is used as the bias voltage (top of Figure 9.12). If the actual ADC output code A does not exceed the reference level B, the code comparator sets the flip-flop to the "1" state, turning on the upper current switch and turning off the lower current. This causes the integrator output to begin to rise.

The integrator output is attenuated and is used essentially as a fine adjustment to the bias voltage. After a few clock inputs, the ADC input eventually reaches a level where the output code exceeds the reference code B. This reverses the integrator output until A no longer exceeds B and the voltage rises again.

At this point, the servo loop "hunts" above and below the statistical threshold with a small sawtooth waveform. (Illustrations are given in Souders' and Flach's article in Chapter 13.) The average of this voltage is read by an integrating voltmeter and is the *upper of the two decision levels* for the reference code B.

A more precise version of the standard loop is shown in Figure 9.13. The bias is provided by a precision programmable reference or DAC and the sawtooth is made very small by attenuation of the integrator output. The difference

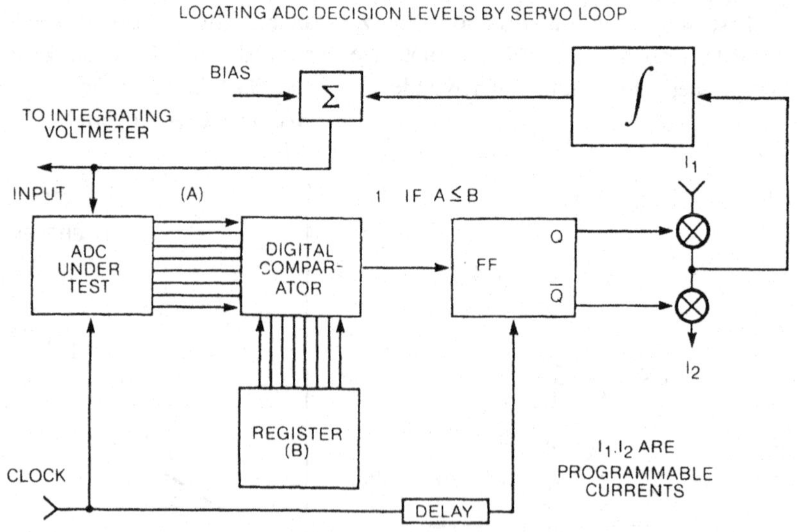

LOCATING ADC DECISION LEVELS BY SERVO LOOP

Figure 9.12: Locating ADC decision levels by servo loop

between the ADC input and the reference voltage is amplified before sending it through the cables of the ATE system.

Probabilistic Estimation of ADC Input Noise

We cannot measure the equivalent input noise of an ADC by the method of Chapter 8, that is, by dynamically short-circuiting the input and measuring the output noise. With quantizers, the input must be activated or "kept alive" in some way.

The servo loop can be modified to find the input noise statistically. Let N2 represent the number of times in some arbitrary time interval that code A exceeds B. Let N1 represent the number of times A is equal to or less than B. When the two reference currents are equal, the servo loop stabilizes the average ADC input voltage such that N1 = N2.

But now suppose the current sources in the servo loop are unbalanced by the ratio 5.3:1, as shown in Figure 9.14. The servo loop will still stabilize the DC level applied to the ADC input but at a new DC level. In order for the average input current into the integrator to be zero, the duty cycle of the flip-flop must be altered by the hunting action so that the average *on* time of the lower switch is 5.3 times greater than the average *on* time of the upper switch.

Assuming a constant clock rate, the ratio, N2/N1, must be 5.3. The probability of getting the higher code is N2/(N1+N2), or 5.3/6.3 = 0.8413. The voltage deviation

from the mean or code center required to do this is the *standard deviation* of the curve. This is the area or cumulative probability of the Gaussian curve up to one sigma (Figure 9.15). The equivalent ADC input noise thus numerically equals the shift in input voltage when the reference currents are unbalanced by a 5.3 ratio. Since the integrator output is attenuated before reaching the ADC, the magnified shift is easily measured prior to attenuation.

Dynamic Testing

There are a number of ways in which ADCs and DACs may be tested under realistic or dynamic conditions. For a DAC, this is usually done by applying a digital pattern of the type and speed typical to the end application and by analyzing those properties of the output response that influence the performance of the system in which the DAC is to be used. Such properties may include only the flat, settled portion (the plateau) of each step or all of the step including glitches and transient phenomena, or the reconstructed, continuous, band-limited output waveform. For an ADC, dynamic testing is performed by applying a suitable analog waveform and by analyzing the response vector (numerical vector, not digital).

To test the DAC used in pulse code modulation (PCM) (compact disk) digital audio, for example, the input pattern might represent a perfectly sampled sinusoid or multitone

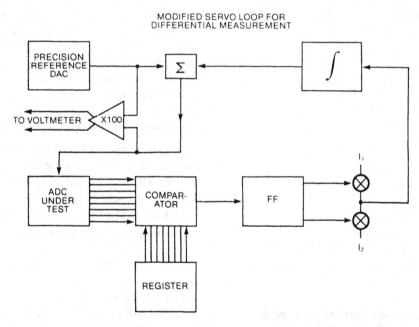

Figure 9.13: Modified servo loop for differential measurement

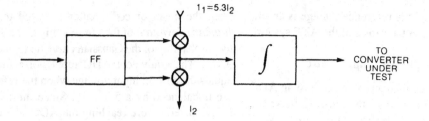

Figure 9.14: Unbalanced integration for noise measurement

waveform. It is not the step or plateau response that is important here, but the *reconstructed waveform*, as seen either through a laboratory-quality reconstruction filter (if the DAC alone is to be evaluated) or through the type of filter to be used in the end application (if the overall performance is to be evaluated). The audio waveform is analyzed for those parameters we would normally measure in audio or video circuits: harmonic distortion, IM distortion, signal-to-noise, noise power ratio, envelope delay distortion, etc.

DAC output analysis can be performed by high-quality analog instruments, or by DSP methods using a digitizer and the FFT operation. If the DAC is of high quality, it may be difficult to find a digitizer capable of providing the speed and/or linearity sufficient to directly sample the DAC output signal without introducing its own errors. There are several methods that combine the best of analog and DSP techniques, however, to look at noise and distortion components down to −110 dB while maintaining the speed and flexibility of DSP. One such fixturing method is described by Landry in an article in Chapter 13 entitled "Production Testing of PCM (Digital) Audio Circuits."

In DAC testing, it is most important to know whether the output is to be tested "raw" (stepwise) or through a de-glitcher, or through a reconstruction filter, or both. The clue is always how the DAC is to be used in its end application. Video and audio DACs, for example, are normally used with reconstruction circuits. Deflection DACs, on the other hand, are generally followed by wideband amplifiers, so the step or pulse shape and level are both critical, not a reconstructed analog waveform. Likewise, if the DAC is followed by a de-glitcher, its transition errors are less important than the plateau properties. A good method for this kind of testing is the Walsh-Rademacher procedure shown by Souders and Flach in the article mentioned earlier.

The following plots show some of the dynamic relationships in DAC and ADC spectral testing. Figure 9.16 shows the result of a simulation of a 16-bit converter, ideal in quantization, but exhibiting classical second and third harmonic distortion (e.g., as in the buffer amp). The total nonharmonic noise is -98.1 dB, but none of the nonharmonic bins shows a component anywhere near this level. This is an example of one of the several techniques made possible by coherent operation, by which digitizer quantization distortion is spread over a large number of bins. The *average bin power* is thus very low, enabling the digitizer to "see" DUT distortion components below the 98 dB level.

Noise Improvement Figure

In an ideal ADC, the quantization distortion occupies only the odd bins. If the quantization is not perfectly symmetrical, distortion occupies all bins. With relatively prime M and N, therefore, quantization noise is spread over K bins, where K is a number between N/4 and N/2.

The noise improvement figure (NIF) in dB is

$$\text{NIF (dB)} = 10 \log K \tag{9.5}$$

In Figure 9.16, N is 1024, so K falls between 256 and 512. The NIF is thus 24 to 27 dB, which means the *average bin power* due to quantization is 24 to 27 dB less than −98

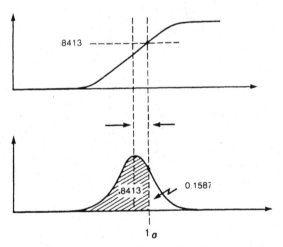

Figure 9.15: Noise distribution (bottom) and cumulative probability distribution (top)

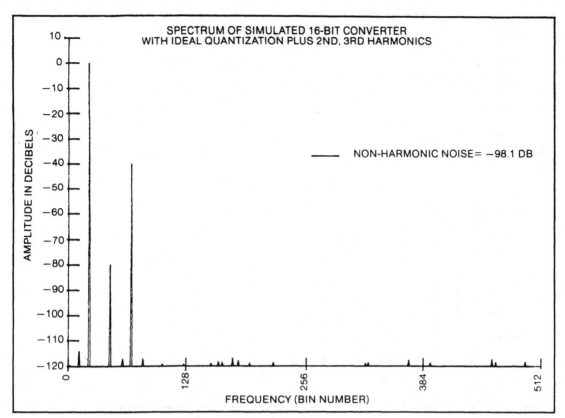

Figure 9.16: Spectrum of simulated 16-bit converter with ideal quantization plus second and third harmonics

dB, or 122 to 125 dB. The *average bin* thus falls below the bottom of the plot (−120 dB).

Quantization distortion tends to be Gaussian-distributed in FFT bins, so the *worst-case bin* is expected to show a power level about 10 or 11 dB higher than the average bin. This is approximated by the relation,

$$\text{worst-bin NIF} = 10 \log K - 10 \text{ dB} \qquad (9.6)$$

based on the smaller (pessimistic) value of K. By this formula, we would not expect any nonharmonic bin to show a power level higher than −112 dB with ideal 16-bit quantization. The computer simulation of Figure 9.16 shows good agreement with this limit.

Random Voltage Noise

Since random (Boltzmann) circuit noise is uncorrelated with other ADC or DAC error components, its power is additive to that of the other components. There is still a 10 dB spread (roughly) between average and worst-bin level. Figure 9.17 shows the previous 16-bit converter with added circuit noise amounting to 10 ppm relative to the sinusoidal test signal (−100 dB).

Exercise 4: Calculate the sum of quantization power and random noise power, and compare with the spectral sum of Figure 9.17.

Induced Jitter Noise

No real ADC is capable of perfectly regular sampling. Likewise, no real DAC has identical transition delay each time it is clocked. And no real timing logic can produce perfectly regular clock spacing. The result is *jitter*, or random clock-to-clock timing errors, causing each ADC sample, or DAC transition, to occur sooner or later than its ideal time. In ADC sampling, the amount of random irregularity is termed the *aperture uncertainty*.

A *histogram* of the timing errors usually shows a roughly Gaussian distribution. The standard deviation of this distribution is the RMS jitter, and the peak-to-peak error is roughly six times this.

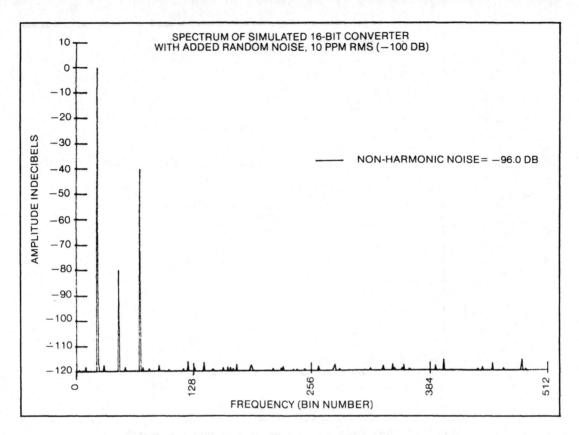

Figure 9.17: Spectrum of simulated 16-bit converter with added random noise, 10 PPM RMS (-100 dB)

When a moving analog signal is sampled at any portion of the waveform with nonzero slope, timing error *induces* a voltage error. With large N the cumulative effect of these voltage errors is essentially the same as that of random, additive, noise voltage. That is, we cannot tell from the spectrum of a quantized sinusoid whether the random noise is due to voltage (Boltzmann's) noise or if it is induced by timing jitter. By running *two* spectral tests with different amplitude sinusoids, however, the difference is easily seen: jitter-induced noise "tracks" with signal amplitude (i.e., varies in direct proportion), whereas voltage noise remains essentially constant.

Jitter-induced noise also tracks with signal frequency, whereas voltage noise, ideally, should not. In high-speed converters, however, there are pseudo-noise effects of imperfect quantization that vary with signal frequency, so changing F_t may not be as reliable in separating jitter from voltage noise as changing amplitude.

The average random power induced by jitter is half that produced at 0 degrees of a sinusoid (the point of maximum slope). From this, the ratio of signal-to-RMS induced jitter can be calculated. In dB, this ratio is

$$\text{induced S/N} = 20 \, \log\left(\frac{\text{signal period}}{\text{RMS jitter}}\right) - 16 \text{ dB} \quad (9.7)$$

Figures 9.18 and 9.19 show the noise in a D/A input pattern created by computing sinusoidal samples at intervals randomly displaced from uniform intervals by a Gaussian variable. In Figure 9.18, the RMS jitter is 10 ppm (1/100,000) of the signal period. In Figure 9.19, the jitter is increased to 100 ppm (1/10,000).

Exercise 5: Compare the expected signal-to-jitter noise given by the above formula with the summed nonharmonic spectral power shown in Figures 9.18 and 9.19.

Exercise 6: Find the RMS jitter in nanoseconds in Figure 9.19 (100 ppm) if $F_t = 10$ kHz.

Exercise 7: Find the RMS jitter in picoseconds for 100 ppm if F_t equals 40 MHz.

Figure 9.20 shows the effect of the same 100 ppm jitter on 12-bit conversion. In this instance, the performance is no better than that of a 12-bit converter with the same jitter.

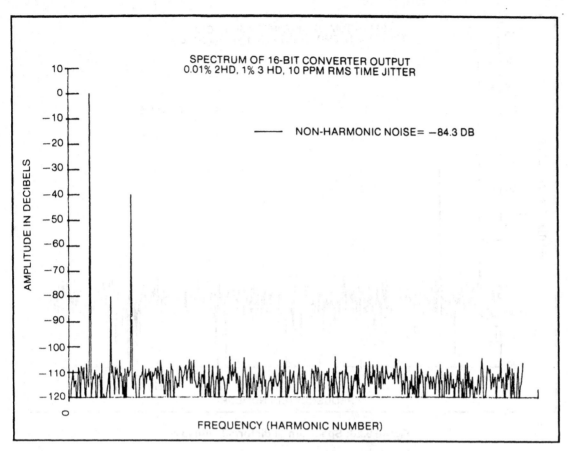

Figure 9.18: Spectrum of 16-bit converter output

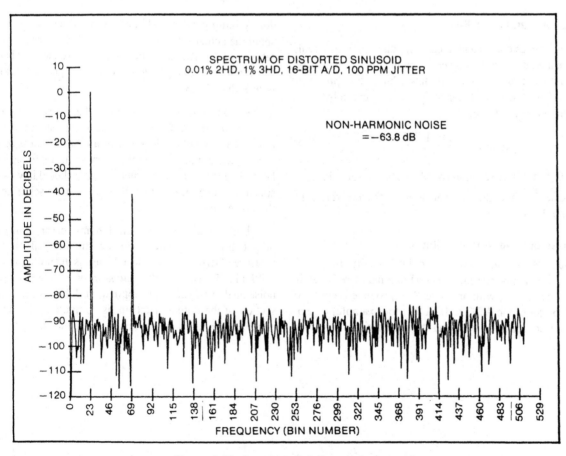

Figure 9.19: Spectrum of distorted sinusoid

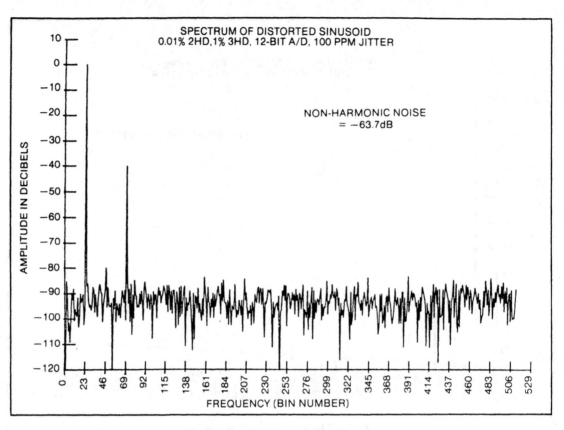

Figure 9.20: Spectrum of distorted sinusoid

Equivalent Number of Bits

It is sometimes convenient to compare the actual, in-circuit performance of converters by equating the distortion and/or noise with an ideal converter with fewer bits. The *equivalent number of bits* based on distortion or noise is given by solving Equation (9.4) for n:

$$n \text{ (equiv)} = \frac{D - 1.761}{6.02} \qquad (9.8)$$

where D is the full-scale signal-to-distortion ratio in dB.

Exercise 8: Find the equivalent n of the converter in Figures 9.19 and 9.20.

Idle Noise and Noise Power Ratio

Because A/D conversion is an intrinsically nonlinear process, we cannot measure input (Boltzmann) noise as in linear circuits (i.e., with an "idle," or grounded input). If the DC offset of the ADC is 1/2 LSB, for example, a decision level occurs at 0 volts, and the output codes will ran-

domly jump between odd and even values, no matter how small the actual input noise is. On the other hand, if the off-set is 0 (or 1 LSB, etc.), the ADC is biased at a code center; no output code changes will occur if the peak-to-peak noise is less than 1 LSB.

The input must be activated or "kept alive" by a signal in order for the ADC code output to reveal the true input noise. The so-called *idle noise* of an ADC is best redefined as the background noise with a very small input signal. The keep-alive stimulus, or "dither" signal, should have a known spectral distribution so that its power can be separated from the noise power.

In high speed linear ADCs, dynamic errors can produce output noise far greater than the small-signal noise. One measure of dynamically induced noise is the noise power ratio (NPR) in dB. To obtain the *numerator*, a wide band pseudo-noise analog waveform is applied, and the power in a small section of the output spectrum is computed. To obtain the *denominator*, a matching notch is cut into the input spectrum to prevent input power from directly appearing in the measurement band, and the "spill-over" or induced power is measured. The proper input level is that which, by experiment, produces the largest ratio.

Separating Quantization Distortion from Noise

In the previous plots, quantization noise resembles random noise in its spectral distribution. This is the case when the FFT period is 1 unit test period (UTP). Since quantization noise is periodic with the UTP, however, extending the FFT over several UTP will cause the quantization distortion power to bunch, leaving other bins relatively free to hold true random noise.

Figure 9.21 shows this bunching in an 8-bit flash converter with jitter-free clocking. By making the apparent M and N

112 and 1024, respectively, the true M and N become 7 and 64. The common factor is 16, meaning that the FFT period spans 16 UTP. Between every pair of quantization bins there are 15 others, empty except for random noise.

In Figure 9.22, these bins have been filled by deliberately adding random noise to the ADC clock timing, such that the RMS jitter is approximately 100 ppm. The total nonharmonic power has not changed appreciably (in fact, it seems to have dropped, but this is an artifact of certain complex changes in correlation of the FFT components and should be ignored here). But the method of bunching has made the two plots quite distinct in appearance, and made it possible for the computer to separate true random noise from a much larger quantization power. A further example is seen at the end of Chapter 10, along with other spectral techniques.

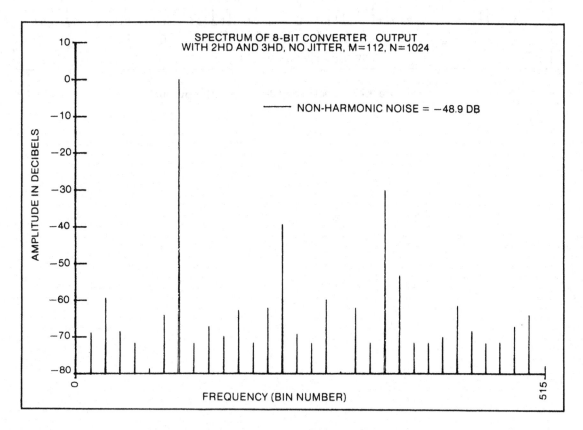

Figure 9.21: Spectrum of 8-bit converter output

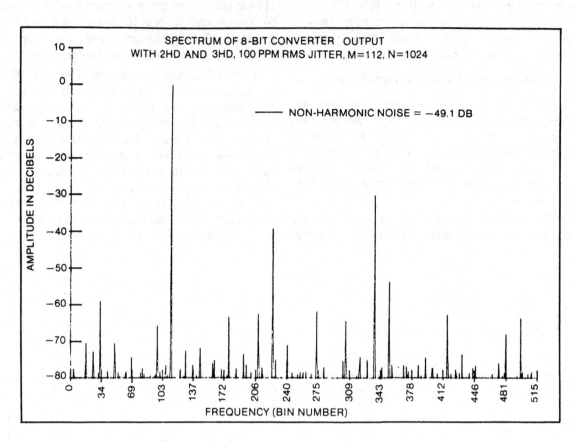

Figure 9.22: Spectrum of 8-bit converter output

Chapter 10
Techniques for Flash
Converter Testing

Chapter 10: Techniques for Flash Converter Testing

Parallel, or "flash," A/D converters (ADCs) present a challenge to the test engineer not only because of the speeds at which they work—20 to 100 Megasamples per second (Ms/s) for most 8-bit units, with a few designs exceeding 250 Ms/s—but because they are *code-dominated converters.* Each step can have an error unrelated to that of any other step. Unlike *bit-dominated converters* (e.g., successive-approximation ADCs), they have no major transitions or carries, and one cannot use short-cut linearity computations. Every code should be tested and used in analysis.

Flash converters are also unlike other ADC types in that it is not clock rate that affects dynamic linearity nearly so much as the frequency and waveshape of the input signal.

The reasons for this and the preceding differences can be seen by comparing the traditional bit-dominated conversion principle (Figure 10.1) with the parallel (flash) principle (Figure 10.2). In the flash ADC, there is an input comparator for each decision level and, often, for the two *virtual levels* as well. An 8-bit flash ADC may thus have 255 to 257 comparators. The outputs of the two virtual-edge comparators (if used) are typically used for over-range indication.

Ideally, the logic state seen by the decoding circuit (255 inputs in an 8-bit unit) should have a clean "horizon," meaning that the logic output of all comparators below the horizon should be "1" and the output of all those above the horizon should be "0," or vice versa. There are only 256 clean-

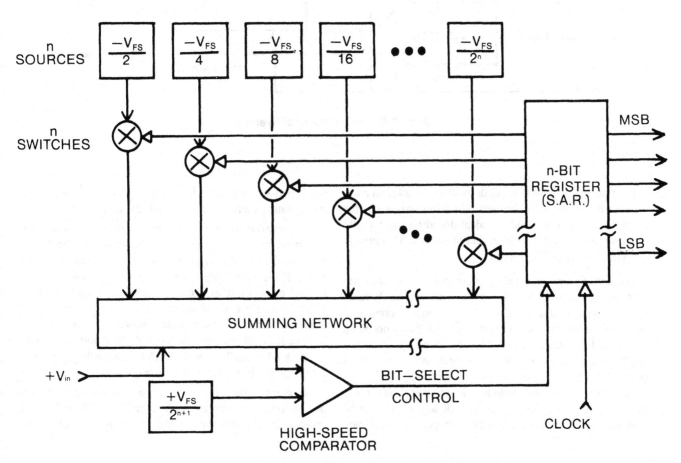

Figure 10.1: Concept of bit-dominated conversion

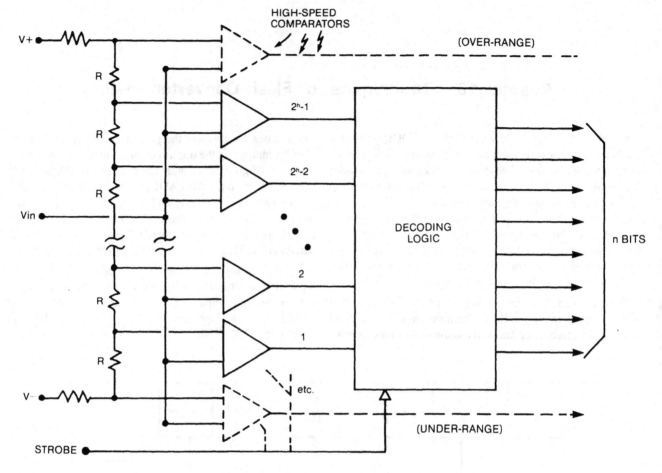

Figure 10.2: Parallel (flash) conversion

horizon states, ranging from all 0s to all 1s. Ideally, all other logic permutations would be constrained (nonexistent by the nature of the system), and the ideal decoding logic would not have to interpret any other kind of pattern. This greatly simplifies the decoding circuit.

Because of propagation delay differences between the comparators, however, fast-moving input signals can easily cause the latched (strobed) comparator state to contain "bubbles," or out-of-place 1s or 0s around the horizon. Early decoding designs did not assume any bubbles, which meant the decoding logic did not know how to interpret comparator states other than clean-horizon states. Interpreting other states greatly complicates the decoding logic, so, even today, many units can accept only one bubble immediately adjacent to the horizon. (The known universe cannot supply a decoding network that can uniquely interpret all 2-to-the-256th-power input states; there are not enough atoms!)

If a constrained state does appear, the decoding network will produce a binary output code unrelated to the best-fit horizon. (With bubbles, the horizon becomes a statistical concept.) If the false code is only a few least significant bits (LSBs) away from its proper location, it is often called a *glitch code*. Since decoding networks tend to be either NOR dominated or NAND dominated, however, many of these false codes tend to contain all 1s or all 0s. Anomalous codes of this kind are called *sparkle codes* because they can cause bright pixels in digital video displays.

The great variety of error sources in flash converters means that no one test method can be considered "best" for all designs. The methods we consider here include the following:

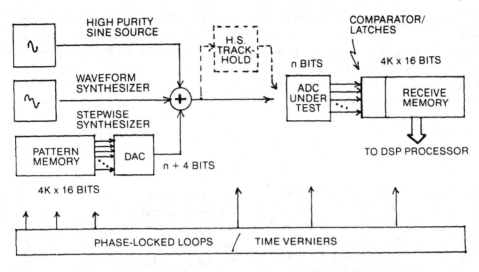

Figure 10.3: Representative high-speed ADC test fixture

1. Linear histograms
2. Weighted code centers
3. MIL-SPEC regression line method
4. Sine wave histogram
5. Spectral analysis
6. Glitch testing by unscrambling
7. Differential gain and phase

All of these tests can be performed with a common type of fixture of the simplified form shown in Figure 10.3. The fixture is assumed to be coherent, and includes a high-purity sinusoidal generator with a frequency range of (say) 50 kHz to 100 MHz. For multitone and pseudo-noise analog inputs, there is a conventional waveform synthesizer, using a high-speed pattern RAM, video DAC, and reconstruction filter. Often, this is a so-called *interpolation synthesizer*, in which the digital pattern is the waveshape *derivative*, and where reconstruction is done by an *integrator*, providing straight-line interpolation between sample points (i.e., first-order hold).

There is another source, intended for step-wise signals (zero-order hold) and DC levels, using a de-glitched DAC, and no filter. These are summed to form a composite signal, if needed. The composite may be fed directly to the ADC under test, or via a reference high-speed track-hold circuit. A good fixture should allow pattern (word) rates to at least 125 MHz for video applications and ideally supply ADC clocking to 300 MHz. When the clock rate exceeds the word rate, every Kth word may be latched into the receive RAM until the complete vector is formed, generally where K is a prime integer. All edges are presumed to have programmable delay.

Linear Histogram Testing

The word "histogram" means a drawing or record of the past. In ADC testing, a histogram shows how many times each different output code word appears in the response vector, without regard to the location. The complete list is derived from the ADC output vector by a single TALLY instruction. Coherence with prime M is assumed in all histograms here.

The analog input can be any wave whose amplitude distribution is known. A linear ramp simplifies the computation, however, because step width is *directly proportional to the tally of each code*. Differential linearity error (DLE) is quickly obtained by subtracting the average step size. Ramp amplitude is not critical but is usually made a little larger than the nominal ADC range to allow for unit-to-unit variation in analog gain.

Let MAT L represent the output vector of an eight-bit ADC, containing (say) 2048 words. MAT K is the complete tally of all 256 unique codes, while MAT T is the working tally of 254 valid (i.e., doubly bounded) steps.

$$MAT\ K = TALLY(L)$$
$$MAT\ T = K(1\ TO\ 254)$$

(10.1)

In MAT K, element K(1) is the number of times code level 1 appears in the list, MAT L; K(2) tells how many times code level 2 appears, and so on. The final entry, K(256), is the tally of the "0" codes. This, like K(255), is discarded because those two steps have no outer bound. MAT T is the *code histogram* of the ADC. Figure 10.4 shows the code histogram of a commercial 8-bit flash converter.

The histogram is only an intermediate step, however. It is the differential linearity error (DLE), that is of interest.

To find the DLE, we subtract the average count from each tally. If the answer is to be expressed in LSBs, the difference vector is divided by the average count.

$$MAT\ A = AVG(T)$$
$$A = A(1)$$
$$MAT\ D = [-A] + T$$
$$*\ MAT\ D = [1/A] * D$$

(10.2)

Figure 10.5 shows a plot of MAT D for this same 8-bit converter. In this example, A = 7.799 counts per step, slightly less than 2048/256, because the ramp is slightly larger than the full-scale range (FSR). The DLE resolution is roughly 1/15 LSB. By itself, this is adequate for most tests. But it is often helpful to collect more ADC samples to reduce

---

* (To save time in production, DLE can be expressed in code counts, not LSBs. This eliminates the division operation.)

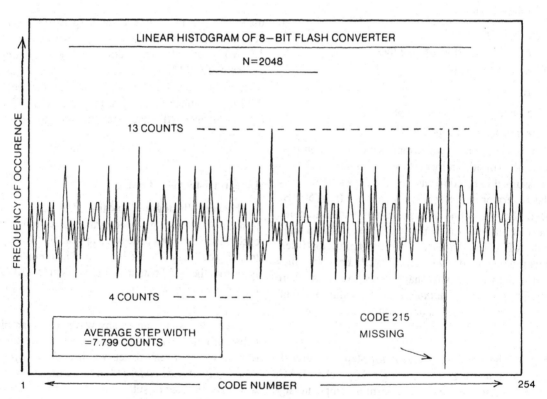

Figure 10.4: Linear histogram of 8-bit flash converter

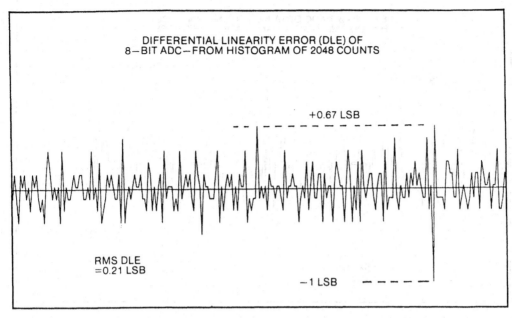

Figure 10.5: Differential linearity error (DLE) of 8-bit ADC

the influence of noise on instantaneous decision levels. 4 K to 8 K vectors will generally do a good job for most 8-bit converters.

Even in a so-called static test, linearity is a function of more than the DC decision levels. Inequality of step widths is partly caused by comparator noise, which randomly displaces the instantaneous decision levels. With vectors of 4 K or greater, there are usually enough samples per step that noise is not important in 8-bit ADCs. But it is always advisable in debugging a program to compare several curves from the same device to see if random noise is a problem.

There is also a mathematically induced error in linearity measurement: the quantization of histogram tally counts. If the actual step width were equivalent to 2.4 counts, for example, the tally would show only the nearest whole number, two counts. The step would wrongly appear to have a DLE of -0.17 LSB.

The solution is simple, of course: Use a slower ramp to increase the tally resolution. You can see the difference by comparing Figures 10.6 and 10.7. Tally quantization is obvi-

ous in the upper DLE plot, derived from 1 K points, but not significant in the lower plot, derived from 4 K points.

Histograms with Sparkle Codes and Missing Codes

Histograms are based on an ADC model that is monotonic, a reasonable assumption for flash converters in static (ramp) testing but a very poor one with fast moving signals. Many flash converters become nonmonotonic with inputs at or near the rated analog full-power bandwidth. Suppose MAT T, the actual ADC output *time vector*, shows two occurrences of code 250 between codes 145 and 146. The *tally* will not reveal the abnormality. In MAT T, the TALLY integer of code 145 is followed immediately by the TALLY integer of code 146. The transfer curve derived from the histogram (only) seems normal.

What happens to the two anomalous counts of 250? All they do is increase T(250) by two counts, making it appear that step 250 is just a little wider than nominal. If the average width is 15 counts, that's just 0.13 LSB and might be perfectly acceptable. For that matter, had the anomalous code

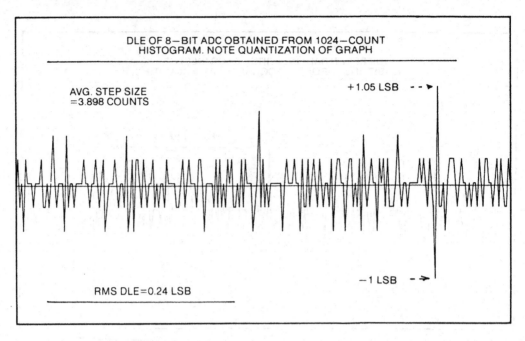

Figure 10.6: DLE of 8-bit ADC obtained from 1024-count histogram

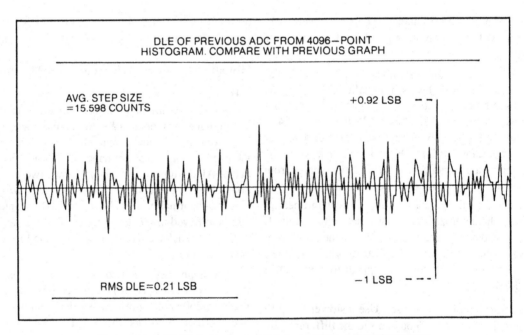

Figure 10.7: DLE of previous ADC from 4096-point histogram

been 255 or 0, not uncommon in "sparkling," it would have been discarded entirely!

The fact that the histogram is blind to nonmonotonicity is seldom mentioned in applications notes. But it is a serious limit in dynamic testing and explains why the proposed MIL-STD method uses a more thorough method that examines the actual positions in which codes occur.

Sparkle and glitch codes do not usually appear in static tests, however. Nonmonotonicity around the code edges because of noise or jitter is commonplace, but this affects only a few adjacent codes and is not serious if vector length N is large. So the linear histogram is a useful way to obtain *static* DLE directly. And, as you can see, it does pick up *missing codes*, wherever the DNL = -1 LSB. Because comparator noise can make a narrow step disappear for one test, or make a missing code *appear*, proposed MIL STD tests define a missing code as any code whose step width is less than 0.15 LSB.

Obtaining the Transfer Function from the Histogram

The TALLY vector provides more than histogram information; it also lists the step widths in the order of their numerical code sequence. If the transfer function is monotonic, this is also the order in which they physically occur with an ascending ramp. MAT T then represents the *derivative function* of the ADC transfer curve.

From MAT T, the transfer curve may therefore be found by *integration*. One way to produce the 254 summations needed for an 8-bit ADC is a simple FOR-NEXT loop:

$$C(1) = <CONSTANT>$$
$$FOR\ I = 2\ TO\ 254$$
$$C(I) = C(I-1) + (T(I)+T(I-1))/2 \qquad (10.3)$$
$$NEXT\ I$$

The result, MAT C, is a list of the actual code centers *relative to the center of step 1*. It is the ADC *transfer function*.

Note that with integration, noise has little effect on $C(I)$. If noise moves the edge separating step I-1 and step I, the computation of $C(I)$ is not affected. If, instead, noise moves one of the outer edges, the error will be only 1/2 of the tally error. Since, in a single vector, noise can increase the width of one step only by decreasing the width of its neighbor, noise errors are not cumulative in the computational loop.

Integral Linearity Error from the Transfer Curve

The constant C1 is analogous to the constant of integration. It should be the center voltage of step 1, but the TALLY does not tell us where this center is. To plot integral linearity error (ILE) from the transfer curve, we generally pretend that step 0 is centered at 0 volts and that C(1) equals the average step width A. Then the *reference*, or linearized, code centers of the ADC under test fall at A, 2A, 3A, ..., 254A.

This list, which we shall denote MAT C2, is a reference against which we can compare MAT C, the list of *actual code centers*. To speed its calculation, prepare in advance the integer series, 1, 2, 3, ..., 254, = MAT I, which shows the code centers for an ADC with a transfer slope of one code level per tally count. For the ADC under test, multiply this by A:

$$MAT\ C2 = [A] * I \qquad (10.4)$$

To find the errors due to integral nonlinearity, subtract the two:

$$MAT\ E = C - C2 \qquad (10.5)$$

MAT E lists the ILE as a function of code level and may be plotted to give the ILE graph. In commercial testing, where only the ILE extremes are specified, the XTRM operation may be applied to MAT E.

ILE Directly from DLE: A Fast Method

If it is not necessary to know the transfer function, the ILE function may be found quickly from the DLE vector (MAT D), by combining some of the previous lines. Let the center of step one serve as a reference; then,

$$E(1) = 0$$
$$FOR\ I = 2\ TO\ 254$$
$$E(I) = E(I-1) + (D(I)+D(I-1))/2 \qquad (10.6)$$
$$NEXT\ I$$

MAT E is the ILE function of the device (Figure 10.8).

Note that MAT D, by this method of derivation, has a vector average of 0. This means that E(254) is always close to 0. If the *absolute ILE* is plotted, the zero-error axis will join the curve at its ends.

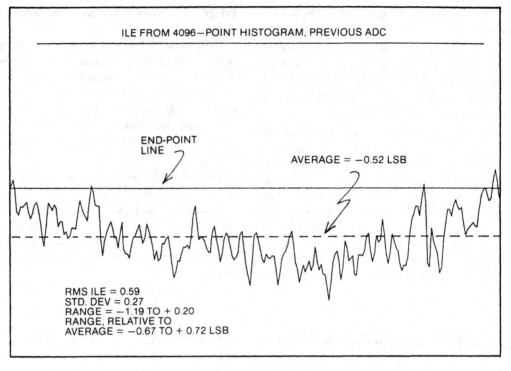

Figure 10.8: ILE from 4096-Point histogram, previous ADC

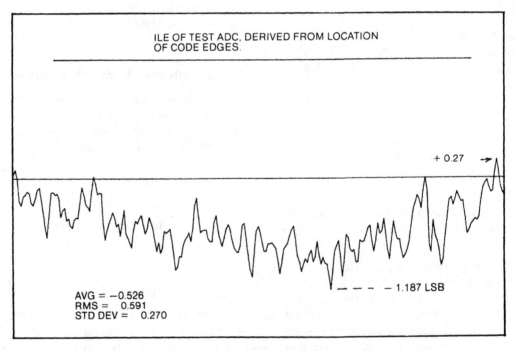

Figure 10.9: ILE of test ADC, derived from location of code edges

Centerlines for Histogram-Derived ILE

The absolute axis is by no means the best-fit straight line when a converter has superposition error—certainly not in a flash converter—where there is no *bit-related error* at all. In Figure 10.8, using the absolute axis makes the peak error seem large (−1.19 LSB), and also the root-mean-square (RMS) error (0.59 LSB).

To minimize the advertised value of ILE, the manufacturer would likely choose a new centerline that splits the peak errors evenly. In this example, the peak-to-peak error is 1.39 LSB, so the new graph would be offset by half of this, or 0.695 LSB. This makes the ILE peaks symmetrical, at ± 0.7 LSB.

If RMS error is important, you might use instead a centerline that is offset by the *average* of the ILE, 0.52 LSB. This reduces the RMS error to the standard deviation, which you can see is only 0.27 LSB, versus the previous value, 0.59 LSB. Mathematically, there is a centerline for RMS error that is even more favorable than the average, but we'll get to that later. For most ADCs, though, the average-derived line is very close to the best fit. To repeat, this is *not* the line that passes through the end points.

How accurate is ILE obtained by progressive integration? For static, linear histograms, it is quite good. Compare Figure 10.8 with Figure 10.9, in which the ILE was laboriously plotted by measuring every individual code edge with a servo loop, essentially eliminating the influence of random noise. In general, if the input ramp comes from a DAC of n + 4 bits (as in Figure 10.8), the results are very similar. Because of its relatively low slope, the linear ramp is considered a static test for flash ADCs, even when F_s is the *maximum rated clock rate of the ADC under test*. For successive approximation ADCs, however, a linear ramp can serve either for static testing or dynamic testing, according to the clock rate.

Integral Linearity from Weighted Code Centers

One disadvantage of integration is that it removes any obvious evidence of a missing code. In effect, it places a missing code directly between the two neighboring codes. By definition, this places the center of this zero-width code right where it would be if the code step had finite width and were perfectly positioned. Moreover, the histogram from which the ILE is derived is blind to out-of-place codes and to noisy decision levels.

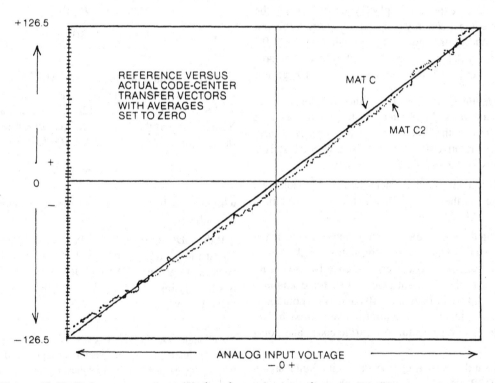

Figure 10.10: Reference versus actual code-center transfer vectors with averages set to zero

All this information is still in the original ADC vector, however. Here is a way to make the information work for us. Let the input be a ramp, rising at a rate slow enough that each step is sampled a number of times. Many of these code counts fall close to decision levels. Since these levels are in constant motion because of comparator noise, a print-out of the code sequence might look like this:

$$\begin{array}{cccccccc|cccccccc} 1 & 2 & 3 & 4 & 5 & 6 & 7 & & 8 & 9 & 10 & 11 & 12 & 13 & 14 & 15 \\ \ldots, & 9, & 9, & 9, & 9, & 10, & 9, & 9, & 10, & 9, & 10, & 10, & 10, & 10, & 10, & 11, & 10, & 11, & 10, & 11 \end{array}$$

To use an earlier analogy, there are "bubbles" in the sequence, only this time, they are made by comparator noise or timing jitter instead of differential comparator delay.

Can you tell by examining the above sequence where the edges and the center of the 10s group are? A simple TALLY will provide an estimate of the step width, but it will not tell you where either the edges or the center are with respect to the rest of the transfer curve. In other words, it will estimate the differential error but not the integral error. By combining this with another simple operation, however, you can estimate the absolute step center as well. Try this with pencil and paper.

First, use an index number to identify the location of the above codes. As an experiment, place your index (1) at the first "9;" for comparison, my text example starts at the first "10." Now, add up the index numbers where each "10" appears. With my indexing, the sum is 71; you should have obtained 107. Let us call that the *weight* of the 10s group.

Now, divide this by the TALLY of the 10s group, which equals 9. My answer is 7.888, and you should have obtained 11.888. The result is the "center of gravity," or *first moment* of the 10s step, relative to the place where we each started counting! This is the *weighted step center*.

In practice, indexing begins where the first decision level is crossed (i.e., at the first "1"), which is also the start of MAT T, the TALLY matrix.

By involving the *weight* of the code group, we have made the statistical center of the step sensitive to the actual or *absolute* positions of each code occurrence. If code 10 also occurs somewhere far to the right—at index 3000, for example—the answer would be dramatically affected! With nonmonotonic conversion, this is a more informative approach than the histogram. It is very useful for sparkle codes and noisy converters.

To implement the weighting procedure with digital signal processing (DSP), let MAT L represent the list of all 4096

output codes generated with a ramp, and assume MAT T (the tally vector) has been previously computed:

$$\text{MAT J = WEIGHT (L)}$$
$$\text{MAT W = J(1 TO 254)} \qquad (10.7)$$
$$* \text{MAT C = W/T}$$

Placing the previous routine into the computer's operating system greatly simplifies the programming and speed. In this example, MAT C is a list of the "centers of gravity" of *all* 254 valid steps of the ADC under test. If you plot them horizontally versus the vertical binary codes, you have the actual transfer curve.

Generally, it is not the transfer curve that is desired, but the ILE curve. To find MAT E, the ILE error vector, we must compare MAT C with a *reference* list MAT C2.

We already determined such a reference list by the vector operation (10.4) and used it to find ILE by integration of the DLE curve. But to make the weighting procedure consistent with the MIL-STD one to follow, let's first remove the DC components from both axes. Graphically, the two code-center lists will be as sketched in Figure 10.10.

The DC component is removed from MAT C in two steps:

$$\text{MAT R = AVG (C)}$$
$$\text{MAT C = [-R(1)] + C} \qquad (10.8)$$

Next, we must do the same to MAT C2. But the result will always be the same for eight bits, namely, the series, $-126.5, -125.5, \ldots, +126.5$. We will label this MAT Y, to represent the Y axis, and generate it prior to running the program. MAT C2, the list of reference centers, is then

$$\text{MAT C2 = [A]* Y} \qquad (10.9)$$

where A, as before, is the average step width (average of the tallies).

If the ADC under test were perfect, its code centers would fall at the positions indicated by MAT C2. Its actual code centers are given by MAT C. If the ADC is monotonic and has no missing codes, the ILE function is the difference of these two vectors.

---

* Although MAT W and MAT T contain only integers, labeling them as floating point (FP) vectors permits rapid division. It also permits the *weight function* to collect numbers far beyond the normal integer range.

$$\text{MAT E} = C - C2 \qquad (10.10)$$

Our initial computation of MAT E for the example 8-bit converter is plotted in Figure 10.11.

Obviously, something is wrong! Unlike the histogram approach, the weighted code-center ILE method is sensitive to missing codes and sparkle codes. It rather dramatically brings them to our attention! In this case, the problem is a missing code, $T(215) = 0$.

In MIL-SPEC testing, ADCs are rejected at this point. But if missing codes are acceptable, the ILE curve can be made usable by placing them where they logically belong, between the neighboring codes:

```
FOR I = 2 TO 254

IF T(I) = 0 THEN C(I) = C(I − 1) + T(I−1)/2      (10.11)

NEXT I
```

Normally, this routine precedes the computation of MAT E, and a record is kept of missing codes. The corrected graph is seen in Figure 10.12.

The MIL-SPEC method, which will be seen next, uses the same weighting procedure and Figure 10.12 is so labeled because it was actually produced by that technique. The only difference is in the reference line, whose slope is not related to the tally average but is that which gives the least mean-squared error. The inverse slope is of 7.79699, when compared with that of the line determined by the *average* A = 7.79919. For this particular converter, there is no appreciable graphical difference.

Note also how close this curve is to the ILE curve obtained by the integration of histogram/DLE data. Figure 10.13 shows what the histogram ILE looks like when replotted on the *average* reference line. Note that the two graphs would differ greatly if there were glitch or sparkle codes. But a missing code is quite different: It does not affect monotonicity.

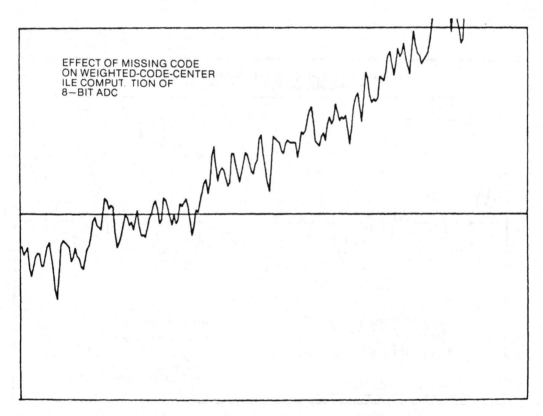

Figure 10.11: Effect of missing code on weighted-code-center ILE computation of 8-bit ADC

MIL-STD Regression Line Approach

The proposed MIL specification for flash converters, MIL-M-38510/132, focuses on integral linearity and also is based on weighted code centers. But the MIL procedure differs from the preceding one in that the reference line is obtained by *linear regression*. Conceptually, the procedure is more complicated than the histogram, but executes almost as quickly with DSP vector mathematics.

The MIL specification shows the classical textbook formulas for regression, but these execute very slowly in the automatic test equipment (ATE) computer. To take advantage of vector mathematics, they have been rearranged algebraically for rapid computation.

Consider Figure 10.10, where the transfer curve has had the DC components removed. The reference line passes directly through the origin.

To find the slope that minimizes the RMS linearity error, we need to compute two things: the unnormalized *variance* of the vertical axis, and the unnormalized *covariance* of the two axes.

The unnormalized variance of any vector is simply the sum of squares. For an 8-bit ADC, MAT Y (the centered Y axis) is always the series: -126.5, -125.5, $\ldots$, $+126.5$. The sum of squares is always the constant 1365568.

The unnormalized covariance (U in this program) is the sum of products of two vectors:

$$MAT\ Z = C * Y$$
$$MAT\ U = SUM\ (Z) \tag{10.12}$$
$$U = U(1)$$

The slope of the regression line equals the *variance* divided by the *covariance*. But it this case, we prefer to find the inverse slope S, which tells us the *number of code counts per reference step*:

$$S = U/1365568 \tag{10.13}$$

Next, project the vertical codes onto the regression line to obtain the list of *reference centers*, MAT C2:

$$MAT\ C2 = [S] * Y \tag{10.14}$$

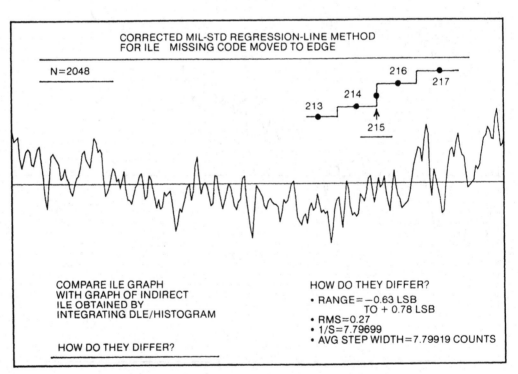

Figure 10.12: Corrected MIL-STD regression-line method for ILE

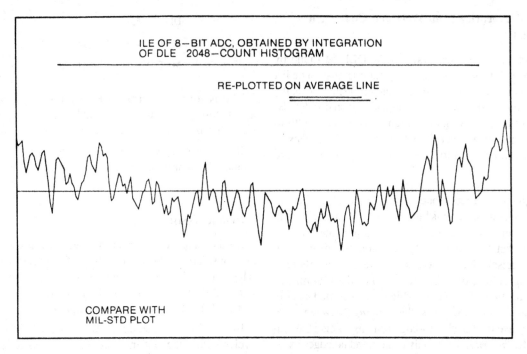

Figure 10.13: ILE of 8-bit ADC, obtained by integration of DLE

The ILE function is the difference of the actual centers and the reference centers:

$$MAT\ E = C - C2 \qquad (10.15)$$

This is what was actually plotted in Figure 10.12.

Extreme Values of Linearity Error

The extremes of ILE in any method are found by the operation,

$$MAT\ X = XTRM\ (E) \qquad (10.16)$$

X(1) gives the positive extreme and X(2) shows which step it is. Likewise, X(3) AND X(4) give the amount and location of the negative extreme. For the RMS linearity error, that is, the excess above the theoretical quantization error,

$$MAT\ R = RMS\ (E) \qquad (10.17)$$

These same operations serve equally to find the RMS and extremes of differential linearity., substituting MAT D for MAT E.

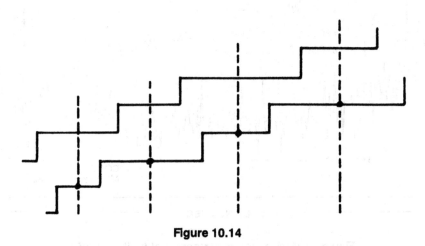

Figure 10.14

Differential Linearity from Weighted Code Center Information

Given weighted code centers, can the DLE be found by differentiation? It sounds reasonable, but it is false. The list of code centers does not contain as much information as the list of contiguous step widths or the list of decision levels. In other words, you can go confidently from step width to centers, but not vice-versa. There is an infinity of transfer curves that can be fitted to any set of code centers (Figure 10.14).

Although code centers alone cannot define an unambiguous ADC transfer curve, it is highly likely that an ADC with uniform center-to-center spacing will exhibit low DLE. It is also true that DLE derived from histogram information does not precisely indicate decision levels where there is glitching, sparkling, or noise. As a result, the preliminary MIL-specification 38510/132, which is concerned chiefly with dynamic behavior, uses the *spacing* between code centers in place of the edge-to-edge spacing. This is analogous to *DAC linearity*, where only centers and no edges exist. The DNL is obtained by progressive differences of the ILE vector:

DIM D(253)

$$\text{MAT D} = \text{E(2 TO 254)} \tag{10.18}$$

MAT D = D − E(1 TO 253)

This is not the identical DLE vector as that obtained from a code histogram, but a comparable measure of step uniformity that perhaps is better for noisy edges.

In the preceding examples, the errors are expressed in code counts. To express them in LSBs, multiply MAT D by 1/S.

Dynamic Testing

Because a single linear ramp changes so slowly, ramp tests are considered to be static tests in MIL specifications, even when the ADC is clocked at rated speed. To obtain dynamic ILE and DLE, an input must be used whose slope is on the order of the slew rate of the comparators.

In principle, one could apply a *triangular wave* with M cycles within the test interval. This increases the slope by a factor of 2M. If M and N are relatively prime, the vector can be numerically shuffled to form an equivalent single-

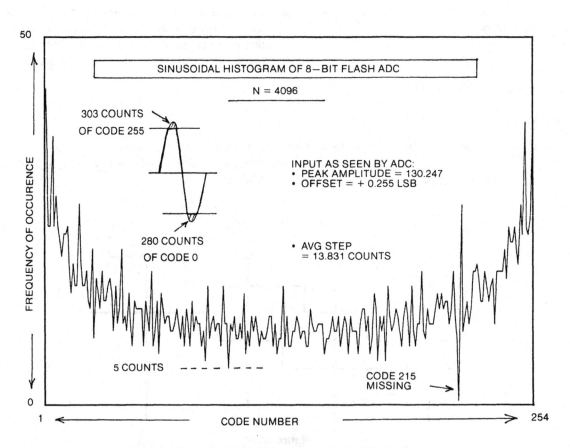

Figure 10.15: Sinusoidal histogram of 8-bit flash ADC

146

ramp sequence (the "primitive" ramp) and the previous histogram or regression techniques applied. This scheme can be used successfully with phase-locked loop (PLL)-based coherent systems with prime-ratio synchronization.

At very high frequencies, however, pure sinusoids are much easier to generate than pure linear triangles, so this multiple-cycle approach is usually restricted to sinusoidal waves. That is, the output vector is formed by allowing the ADC to sample exactly M cycles of a high-frequency sinusoid, where M is relatively prime to N. In this way, the N samples will be uniformly distributed over a single *equivalent cycle* (the primitive cycle) and will fall at N essentially unique points. Of course, the analysis is more complicated than in the linear case.

Sinusoidal Histogram Testing

Figure 10.15 shows the histogram obtained by applying a sinusoid to the input of the previous ADC. Just as with the ramp method, the input waveform is slightly larger that the FSR of the converter, to ensure that all valid steps are exercised from edge to edge. In the example program, the tally vector is labeled T1 to distinguish it from the previous linear tally vector T.

The histogram is not affected mathematically by the value of M, so long as the M/N ratio is irreducible. Changing M merely reshuffles the vector elements but does not alter them. It is analogous to shuffling a deck of playing cards. MAT T1 is not affected.

Although the histogram is not affected mathematically by the choice of M, it is nonetheless physically altered in that the dynamic linearity of the ADC becomes worse and worse as the input slew rate is increased. For fixed N, the maximum rate-of-change of the sinusoid is directly proportional to M. One important reason for using a coherent or prime-ratio-locked test system is that it allows the user to control the dynamic conditions by controlling M, without changing the UTP.

A second advantage of coherence is that it avoids capturing a fraction of a sine cycle, which is very important in histogram testing. N can thus be much smaller than the very large numbers (e.g., 32 to 512 K) indicated in some articles on confidence limits (Ref. [19], A/D reference section in Chapter 14). Such numbers typically pertain to noncoherent fixtures where integer M cannot be assured.

One limitation of the sinusoidal histogram is that steps near the center of the input range receive less resolution than the average. As a result, N must be generally 2 to 4 times that for a good linear histogram. Note that this does not require more electrical resolution in the input ramp but requires more *mathematical resolution*, which is achieved by new values of M and N. For production, N = 4096 is usually adequate but it is better to use 8 K to 16 K for characterizing an 8-bit ADC. If it were not for prime-ratio synchronization, the numbers would be far larger.

A second problem is that the computation of DLE is more difficult with a sine-wave histogram than with a linear histogram. Whereas each step in a linear histogram can be compared against a single *scalar reference*, there is a different number for each of the 254 steps in sinusoidal testing. The sinusoidal reference is thus a numerical vector. In the following example, this vector is labeled T2.

HOW TO FIND AMPLITUDE AND OFFSET OF ADC
INPUT SINUSOID, RELATIVE TO TRANSFER
CURVE OF INDIVIDUAL CONVERTER.

1. N =NUMBER OF COUNTS, INCLUDING END CODES
2. N1=NUMBER OF MAXIMUM CODE COUNTS
3. N2=NUMBER OF ZERO CODE COUNTS
4. C1 =COS (180 * N1/N)
5. C2 =COS (180 * N2/N)
6. SIGNAL OFFSET=V0 LSB's

$$V0 = 127 * (C2-C1)/(C2+C1)$$

7. SIGNAL AMPLITUDE=V LSB's

$$V = (127 - V0)/C1$$

NOTE: CONVERTER OFFSET = −(SIGNAL OFFSET)

Figure 10.16: How to find amplitude and offset of ADC input sinusoid

MAT T2 must be recomputed for each unit under test because the tally near the extremes is very sensitive to the ratio of the signal amplitude to the FSR of the unit under test and also to any offset. To calculate T2, one must first calculate both the offset and the amplitude of the *best-fit sinusoid*. This is the input sinusoid as seen through the "eyes" of the ADC under test.

For sinusoids that are clipped (as in this case), the offset and amplitude of the best-fit sinusoid can be found to useful accuracy by the trigometric procedure shown in Figure 10.16.

The example ADC was tested by applying a sine wave of exactly 130 LSB amplitude with respect to the nominal or design center step size for this family of devices. It was exactly centered with respect to the nominal range of the device (i.e., its offset was zero).

According to the formulas of Figure 10.16, however, this particular ADC "sees" a signal of 130.247 LSB, offset by +0.255 LSB. It thus has an analog gain error of 0.0165 dB. This does not sound like much, but it is most important in calculating the vector T2. Whenever you encounter any formulas for histogram analysis that call for the signal amplitude, remember that this must be the amplitude *as seen by the individual ADC!*

Using the Tally to Find MAT T2

A simple but inexact way to find the reference tally for a sinusoid is to quantize a theoretical sine wave of 130.247 units amplitude and 0.255 units offset. You can create this with a FOR-NEXT loop and quantize it simply by *labeling it as an integer vector.* The result is shown in Figure 10.17.

To find the DLE, subtract the vector thus formed from MAT T1, the actual device tally. The difference vector is then normalized to account for nonuniform distribution of reference tally counts caused by a sinusoid:

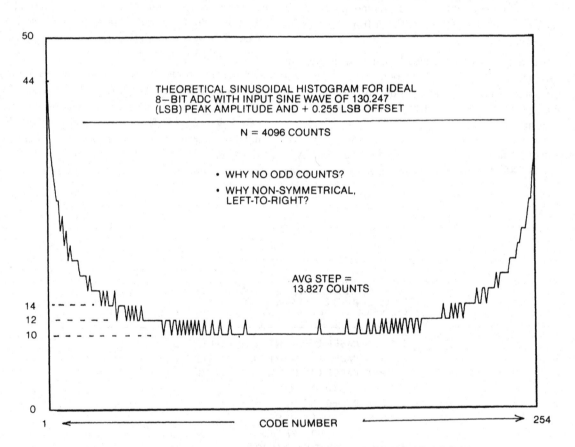

Figure 10.17: Theoretical sinusoidal histogram for ideal 8-bit ADC

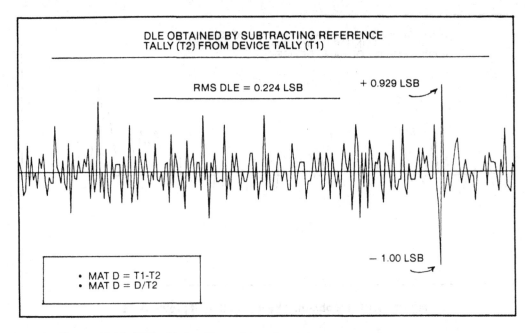

Figure 10.18: DLE obtained by subtracting reference tally from device tally

$$MAT\ D = T1 - T2 \qquad (10.19)$$

$$MAT\ D = D\ /\ T2$$

Unlike the scalar division step in the linear case, this division *is not optional*. It must be performed because T2 is a vector, not a scalar constant.

MAT D, thus linearized, is the dynamic DLE of the device. This is plotted in Figure 10.18. As mentioned, however, the result is not exact. Let us see why.

Errors with Sinusoidal Histograms

Naturally, you expect the dynamic curves of any real device to look different from the static curves. But never accept or fail a device until you are positive that the differences are legitimate and not an artifact of the method. Unfortunately, such artifacts are probably the rule in flash converter testing rather than the exception.

To show what can go wrong, consider what happens with small M, which is to say, a low input frequency. Under these conditions, the DNL obtained by the sinusoidal method should be essentially the same as the DNL obtained by the ramp method. Figures 10.19 and 10.20 show that this is not the case, this time. The blame does not lie with the ADC but with inadequate mathematical procedures.

Our first clue is that the graph tilts upward, not ending at the reference line, as it should with ILE by progressive integration. It implies that the sum of the DLE vector is non-zero. Yet, ideally, the method of computing DLE should have caused the sum to be zero.

What has happened here is a commonplace error when subtracting sets of numbers that are nearly equal in size, as is the case for MAT T1 and MAT T2. Small errors in measurement and/or in roundoff, etc., are highly magnified.

But this is easily fixed: Before computing the ILE, simply take the average of MAT D and subtract it from MAT D. This forces the sum to zero. Integration now produces an ILE curve consistent with the previous types, provided that there are no other errors. To see if there are, we may plot MAT D at this time and compare with the other plots of ILE.

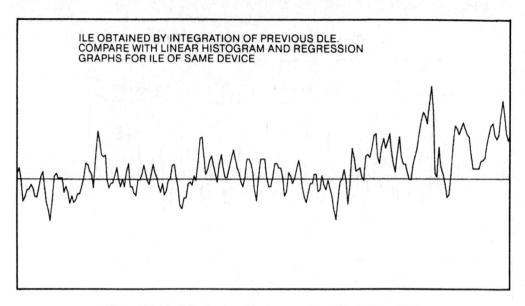

Figure 10.19: ILE obtained by integration of previous DLE

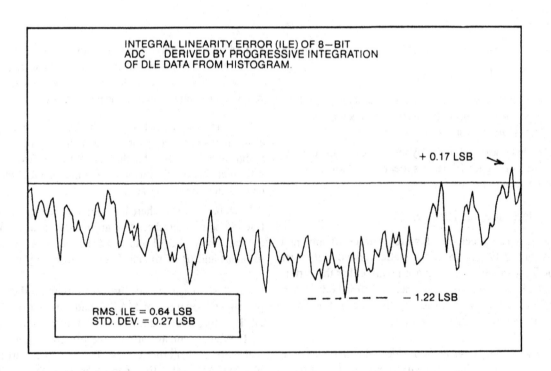

Figure 10.20: Integral linearity error (ILE) of 8-bit ADC

The result of this correction is shown in Figure 10.21. The curve is obviously improved but still is not identical. So far, we have dealt with only one of three significant procedural errors.

Reference Quantization

The second error source is the overly simple procedure that was used to compute MAT T2, in which an integer tally vector was obtained.

Recall that with linear histograms, the average step size (i.e., the average tally) is not an integer but a floating-point number. Well, that holds true for any input waveshape, not just for a ramp. By taking the lazy way to find MAT T2—i.e., using the TALLY command—we caused the reference tally to be *quantized* instead of being a list of floating-point reference widths. It is a natural consequence of trying to measure variable quantities with small integers.

The right procedure is one that provides a continuous, non-quantized reference tally vector. Call that T3. To find this, we need two previously calculated voltages as "seen" by the ADC: the signal amplitude, V, and the offset, V0. MAT T3 is then derived from the probability distribution of a clipped and offset sinusoid where V > (127 + magnitude [V0]).

FOR I = 1 TO 254
T3(I) = (N / 180) * (ASN((I-127-V0)/V)
       − ASN((I-128-V0)/V))   (10.20)
NEXT I

Vector T3 resembles the earlier reference vector, T2, except that it is not quantized. T3 is the limit of what you would get by tallying KN output codes, and dividing MAT T2 by [K], as K is made increasingly large. Division, of course, produces floating-point vector elements. In the limit, this gives the continuous *probability distribution* of a clipped sinusoid of amplitude V and offset V0. This continuous curve is the *best-fit reference* curve for sinusoidal testing (Figure 10.22).

Following this curve is the new ILE plot based on this continuous curve (Figure 10.23). It shows a clearly visible improvement over the quantized reference.

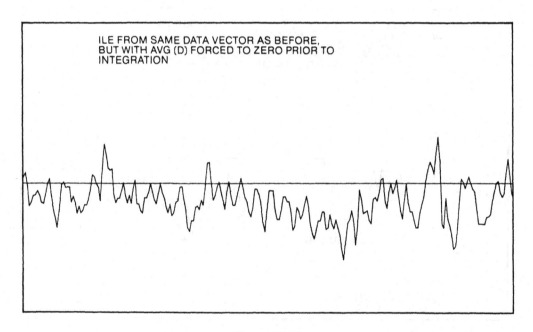

ILE FROM SAME DATA VECTOR AS BEFORE, BUT WITH AVG (D) FORCED TO ZERO PRIOR TO INTEGRATION

Figure 10.21

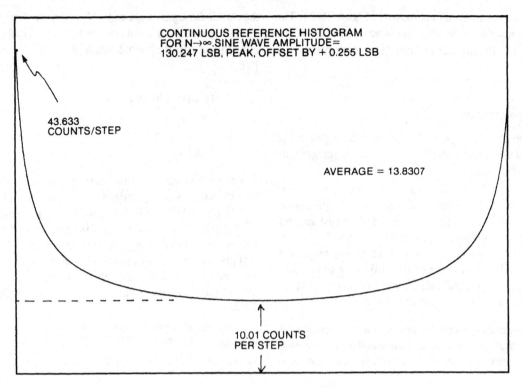

Figure 10.22: Continuous reference histogram

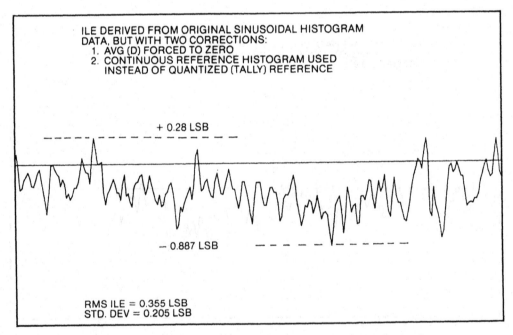

Figure 10.23: ILE derived from original sinusoidal histogram data

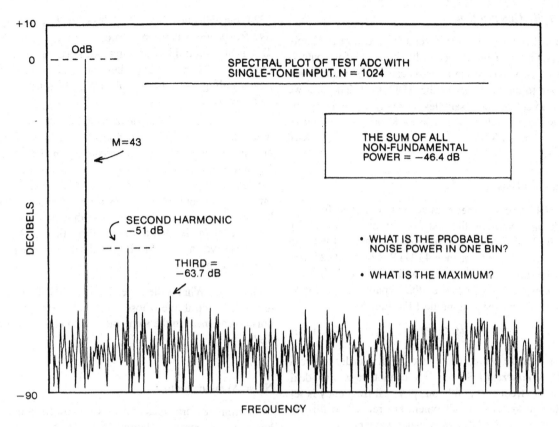

Figure 10.24: Spectral plot of test ADC with single-tone input

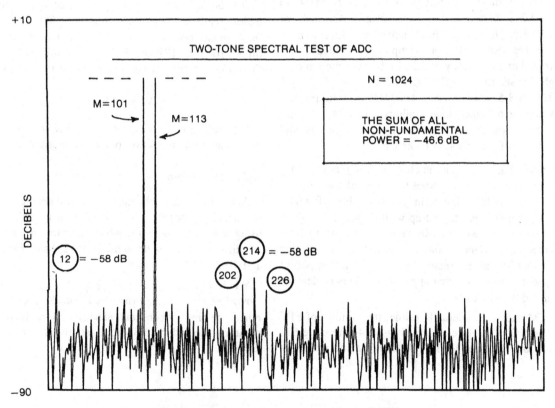

Figure 10.25: Two-tone spectral test of ADC

The third limit to ILE measurement precision is simply the quantization of the device tally, itself. Unfortunately, there is no easy way around this, because this quantization is intrinsic to the operation of the ADC itself. The best we can do is to collect more samples, as many as we have time to process: for production at least 4 K samples should be collected from an 8-bit ADC; for characterization, at least 8 K or 16 K.

Spectral Analysis

The ADC response spectrum is easily obtained from the ADC output vector by the fast Fourier transform (FFT). Figure 10.24 shows the spectrum from a single "tone," or pure sinusoid, at the frequency 43/UTP. Figure 10.25 used a two-tone input with M1 = 101 and M2 = 113. If MAT X represents the times series (i.e., the output vector) the spectrum is found by the magnitude FFT function:

$$\text{MAT F} = \text{MAG(FFT(X))} \qquad (10.21)$$

Frequencies are not marked on the graph in order to stress the fact that *coherent* DSP software cannot, must not, deal with absolute frequency. Only one physical frequency is set in a coherent system and all others are related to this by integer ratios. DSP syntax uses these integers.

That physical frequency is usually the sampling rate, F_s. In the majority of plots in this chapter, F_s is 20 MHz. In Figure 10.24, therefore, F_t is 20 MHz * 43/1024, or 839843.75 Hz. The absolute frequency is not critical in most cases, but the ratio is. If, for example, we needed to test at the color burst frequency of 3.58 MHz, then the multiple M might be set to 183. This gives a frequency of 20 * 183/1024, or 3.57421875 MHz. The DSP program instructions would use only the whole numbers, 43, 1024, 183, etc. It is up to the programmer to set the physical clock rate, and to keep track of the absolute F_s and F_t.

If the input has to be exact in absolute frequency, then it would be F_t, not F_s, that becomes the physical standard. F_s would end up slightly higher (in this case) than 20 MHz. The critical thing is *not* to end up with a fraction of a test cycle left over. You will find, by the way, that most television, video, audio, and communication standards involve numbers that are integer related by specification to permit exact synchronization in other applications. This is ideal for coherent DSP techniques.

When M and N are mutually prime, the number of FFT "bins" is N/2 + 1. Here, where n = 1024, there are 512 frequency bins plus one for DC. At 20 MHz and with N =

1024, the collection interval (UTP) is 51.2 μs, and the primitive frequency Δ is the reciprocal: a little over 19.5 kHz. This is the spectral line spacing. All noise (quantization and random) is distributed over these bins, so the *average* noise power falling into 1 bin is only 1/513 the total, a reduction of about 27 dB.

This means the FFT technique can look "deeper" into the noise than is possible with noncoherent instruments. The amount of improvement in sensitivity in dB is the *noise improvement figure* previously defined in Chapter 9. Here, we see the total noise level is −46.4 dB, but the average bin noise is about −73 dB. As a rule of thumb, the *worst* bin receives about 10 dB more than this, or about −63 dB. That means the second harmonic distortion (at M = 86) is fairly trustworthy at −51 dB. The third harmonic distortion, at −63, is not.

Question: What is the expected range of calculated amplitude of a spectral line whose true level is −63 dB when the noise component in the same bin is also at the level of −63 dB?

Noise Measurement

Spectral measurements of ADCs are obviously useful when harmonic or intermodulation (IM) distortion is important, but they provide other advantages as well.

One is the ability to measure device noise, accurately, under realistic dynamic conditions. To find the nonfundamental power in the previous plots, for example, set bin 43 (or bins 101 and 113 in the 2-tone case) to 0, and then take the sum of squares of the magnitude spectrum:

$$\text{MAT S} = \text{SSQ(F)}' \qquad (10.22)$$

The scalar answer is the total noise power, which may then be normalized to the power of the fundamental component(s).

Noise Separation

Another special application of spectral testing is in separating noise components. This is done in a double test, first using prime M, and then an adjacent submultiple of N. Notice the difference in results in the two graphs of Figures 10.26 and 10.27.

When $N_{(FFT)} = K * N_{(UTP)}$, the *quantization noise* "bunches," appearing primarily in bins K, 2K, 3K, etc. The remaining bins contain *random noise power*, chiefly the comparator noise and the induced noise from timing uncertainty and jitter.

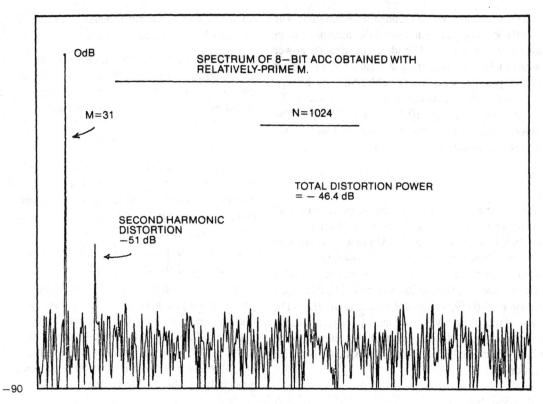

Figure 10.26: Spectrum of 8-bit ADC obtained with relatively-prime M

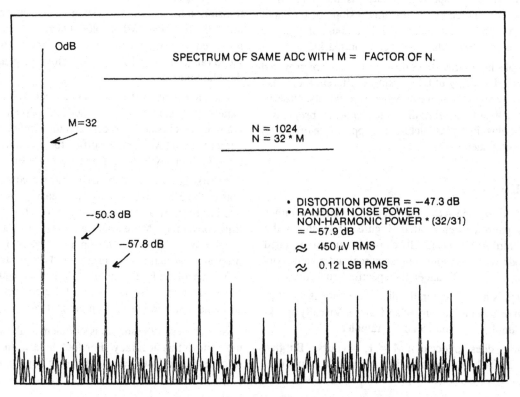

Figure 10.27: Spectrum of same ADC with M = factor of N

The quantization noise power is computed separately. The *excess* over the ideal quantization noise is the *distortion power due to dynamic nonlinearity*. The ideal quantization power is 1/12 square LSBs. In this example, assume that the input sinusoid has a peak-to-peak value of 256 LSBs. Its average (wrongly called "RMS") power is 8192 square LSBs. To check on the arithmetic, take 10 * log (8192/(1/12)). The answer is 49.92 dB, which agrees with the formula for full-scale signal-to-quantization noise for n bits:

$$\text{number of dB} = 6.02 \text{ n} + 1.76 \qquad (10.23)$$

In this experiment, the total distortion power from all sources is roughly 0.1525 units, nearly twice the theoretical value (1/12). Of this, only about 0.01328 units falls between bins K, 2K, etc.; this is the *random noise power*, down about 57.9 dB. Irregular quantization (linearity error) accounts for the remainder, about 0.1392 units. Subtract 1/12 from this. *The difference, 0.0559, is caused by dynamic ILE*. The *square root* of this is the RMS noise due to integral linearity error: approximately 0.24 LSB. Compare this value with the independently calculated RMS error of the MIL-derived ILE plot in Figure 10.12.

Progressive Spectra

One interesting type of spectral test involves progressive spectra: A series of spectral pictures is taken as M is increased. This increases the maximum slope of the input sinusoid from test to test, increasing the amount of dynamic error and eventually inducing glitches or sparkles.

The graphs in Figures 10.28 through 10.30 were provided by Michael McCaffrey of LTX, using a different 8-bit flash ADC than the unit used in the previous experiments. Observe the increase in noise level as the input frequency progresses to the right plus, in the last picture, the appearance of several spurious components.

Unscrambling

With prime M, the ADC time vector produces a rather disorderly pattern when plotted. Figure 10.31 shows the result of a third ADC, specifically chosen because it exhibited a glitch code when its input frequency reached a value corresponding to M = 93 under the specified conditions.

The glitch is not evident in the direct plot, nor is the clipping at top or bottom, nor any other feature. Visually speaking, the samples are said to be "scrambled."

Scrambling does not occur if M = 1, of course. But that is no help in flash converter testing, because it would dra-

matically alter the dynamic behavior of the ADC under test. The glitch code would not appear, for example.

Here is an instance in which the vector unscrambling technique introduced in Chapter 4 is valuable. If X represents the raw ADC vector, then an unscrambled vector Y is produced by the DSP STEP command:

$$\text{MAT Y = X STEP [I]} \qquad (10.24)$$

Integer I is found by the algorithm of Chapter 4. For M = 93 and N = 512, I = 501. The result is the so-called *primitive wave*, in which the time sequence resembles that produced when M = 1, but contains the dynamic errors for the condition M = 93.

Vector Y, the primitive wave, is plotted in Figure 10.32. The glitch and the clipping are clearly seen.

We must be careful not to be fooled by the relative simplicity of the primitive wave. In this case, with N only 512, there are many missing codes because of the sinusoidal distribution. This particular wave does not examine all steps of the ADC. One should always use a histogram (Figure 10.33) to see if N is sufficiently large to examine all codes, especially in glitch code testing.

Differential Phase (DP)

An important consideration in television signal transmission—and thus in testing video converters—is how much the 3.579545 MHz color subcarrier phase shifts as it is carried by a larger signal throughout the luminance or gray video range (Figure 10.34). The reference phase is taken at the blanking level.

The subcarrier test signal has a peak value of approximately 12.5 percent or full scale (25 percent pk-pk). The blanking level is at 75 percent and the white level is at 12.5 percent. Two levels might suffice for a production test, but three or four levels give a better picture of A/D linearity.

At first glance, it might seem that we could apply a DC bias to the 3.58 MHz test signal, measure the phase, apply another bias, measure again, and so on. First of all, this is time consuming. More important, though, in a practical test system, it is unwise to interrupt the stream of samples with program instructions because it tends to destroy the phase coherence of the FFT sample set. It is often required to measure differential phase to +0.1 or +0.2 degrees, and it is best to avoid interrupting the sampling in any way.

At each step, 128 samples is sufficient to allow phase measurement to this accuracy. For four levels, one approach is to collect 1024 samples as shown in Figure 10.35.

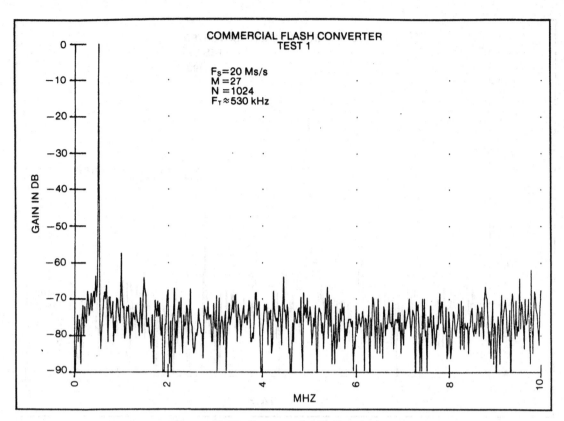

Figure 10.28: Commercial flash converter

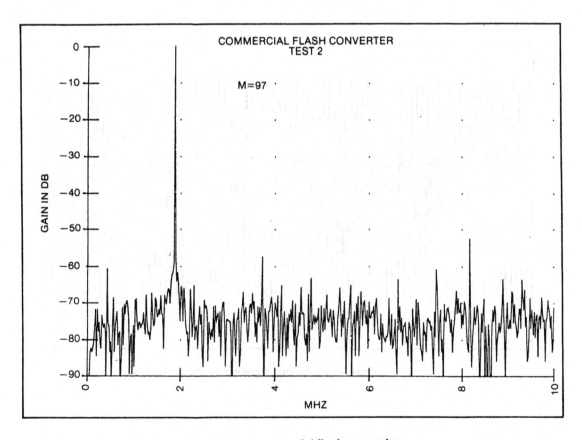

Figure 10.29: Commercial flash converter

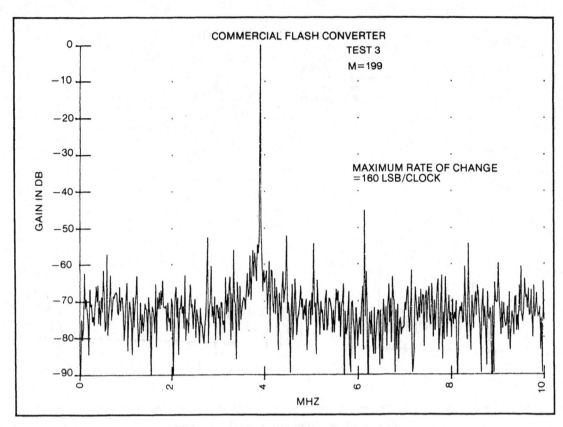

Figure 10.30: Commercial flash converter

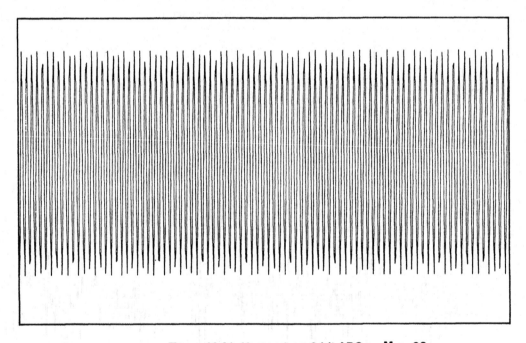

Figure 10.31: Vector from 8-bit ADC;   M = 93

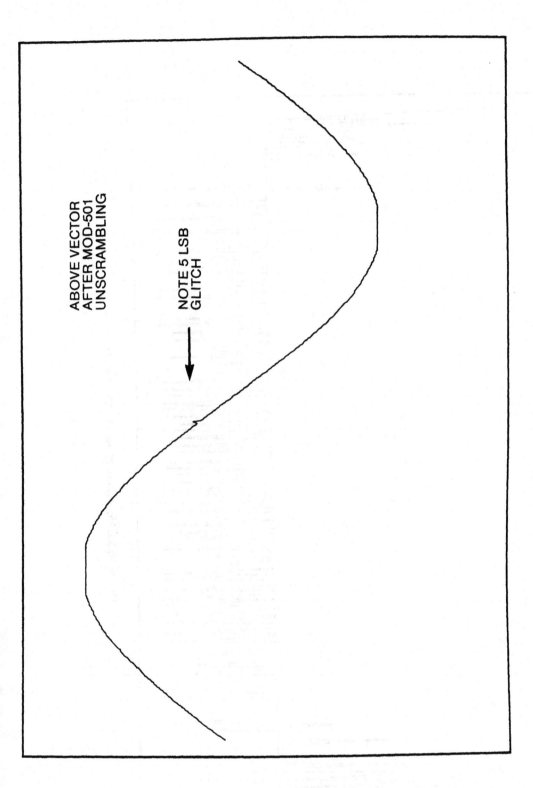

ABOVE VECTOR
AFTER MOD-501
UNSCRAMBLING

NOTE 5 LSB
GLITCH

Figure 10.32: Vector after MOD-I unscrambling

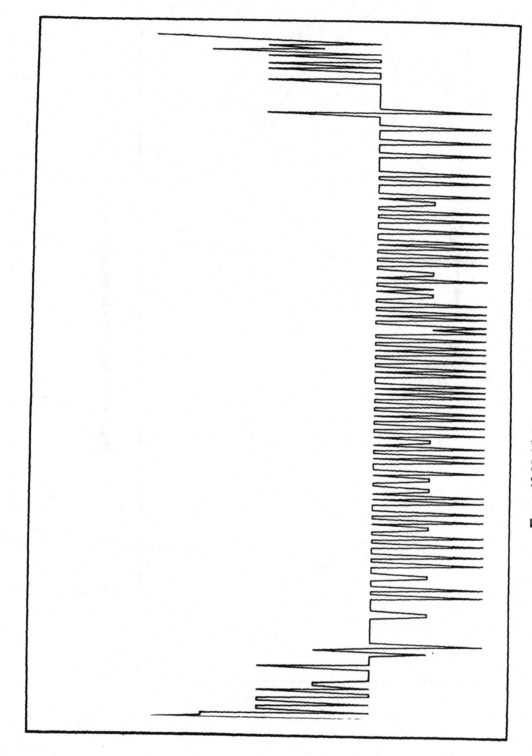

Figure 10.33: Histogram of vector from Figure 10.32

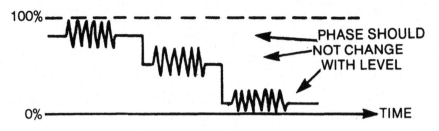

Figure 10.34

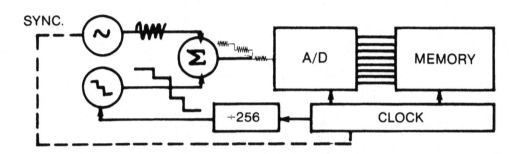

Figure 10.35

A staircase is added to the test signal, with a new level every 256 samples. You can see that many samples are damaged at the step edges because of the finite fall time, overshoot, and other transient errors. We do not care, however, because we will take only the center 128 samples of each step for the FFT.

$$DIM\ B(128),\ A(1024) \qquad (10.25)$$

Taking four coherent FFTs is easy. The software that allows us to do this takes the form

$$MAT\ B = FFT\ (A\ (60\ TO\ 187)) \qquad (10.26)$$

The two numbers after A define the locations in *time* of the first and last sample of an arbitrary group. The numbers 61 to 188 would work just as well, because we have ensured that the test frequency is periodic over 128 samples. The exact starting point is not important, just the period.

To maintain coherence, however, the next three sample sets must be separated by exactly 256 samples. If we started with 60 to 187, then the next sets would be indexed as follows:

$$316\ TO\ 443$$
$$572\ TO\ 699 \qquad (10.27)$$
$$828\ TO\ 955$$

We have not interrupted or shifted the sampling action in any way, except for the differential phase shift in the A/D under test, which is what we wish to measure.

Differential Gain (DG)

In video conversion, it is also important that small signal gain be unaffected by bias levels. The procedure just illustrated simultaneously measures DG along with phase, since the test frequency is the same for both parameters.

For more information on differential phase and gain, see CCIR Recommendation 421.2.

Chapter 11
Incremental Models for DSP-Based Testing with Applications to Transient and Flutter Measurement

Chapter 11:   Incremental Models for DSP-Based Testing with Applications to Transient and Flutter Measurement

Introduction

Using mathematical routines to replace conventional analog instruments is a fundamental concept of digital signal processing (DSP)-based testing. One of the goals of this process is increased test speed, but this does not come about simply because of "DSP." In fact, if DSP routines are performed only by the scalar arithmetic circuits of conventional minicomputers, many common instrument models run more slowly than the analog circuits they are intended to replace.

In practice, high speed modeling is largely a result of using *vector* processing. The algorithms do not deal with individual numbers (scalars) but with numerical strings in which the elements act as coordinates of a single conceptual quantity, or "vector."

In digital testing, each vector element is a binary digit (1 or 0). In analog and mixed-signal testing, each element is a floating-point (FP) number. A vector of this latter type, called a *numerical* vector, is thus of a higher order than a digital vector. In this tutorial, all vectors are numerical unless otherwise indicated.

Vector mathematics enables rapid DSP computation because it can be executed in arrays, using parallel, pipeline processing (i.e., by an array processor). Although speed is the primary goal, the fact that number strings are seen by the programmer only as entities brings an added advantage of programming simplicity.

Software models that employ parallel/pipeline mathematics at every intermediate step are often termed vector models. Ideally, we would like to be able to simulate every laboratory-standard or "reference" analog circuit by a vector model. Unfortunately, there are a number of standard instruments for which this is not possible.

Bench instruments used in wow and flutter testing of cassette decks fall into this category, as do certain other circuits used in the analysis of transients and signals from non-stationary processes. In this chapter, we will examine some of the reasons and investigate an alternative modeling approach based on incremental computations. The intent of this particular scheme is to provide a few simple, highly regular, "generic" procedures that can be performed at machine level by dedicated *scalar* coprocessor modules. Executed this way, these routines can still run faster than equivalent analog instruments.

Limitations of Vector Processors

There are many algorithms that look simple on paper, yet cannot be handled by commercial array processors. In theory, all DSP functions can be expressed by matrices of input-output vectors and, consequently, performed by hardware look-up tables, providing very high speed mathematics. For practical vector lengths, however, there are simply too many logic states to make look-up practical.

Commercial processors attack the problem, first by restricting the class of allowable analytical procedures and then by breaking these down into simple repetitive routines that can be done by hardware or firmware. For the most part, the allowable procedures are sums-of-products expressions, such as are encountered in correlation, the discrete Fourier transform (DFT), fast Fourier transform (FFT), and root-mean-square (RMS) computations. Typically, one stage of the processor pipeline has the job of multiplying vector elements by a scalar constant. A second stage multiplies them by variables (e.g., the elements of another vector). A third stage may add the products, while a fourth stage divides the sum by a scalar quantity. To find the vector average, for example, the multiplying stages are bypassed.

In most stages, there will be a number of identical, parallel operations taking place. But this is possible only if all elements of the input vector are statistically independent and if a uniform mathematical operation is to be applied across the length of the vector at each stage of the pipeline. For example, in the operation,

$$\text{MAT } C = A + B \qquad (11.1)$$

the machine will perform the addition $A(1) + B(1)$ concurrently with $A(2) + A(2)$, $A(3) + A(3)$, etc. No individual

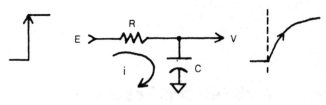

Figure 11.1

operation depends on the values of neighboring elements nor waits upon the outcome of neighboring calculations.

But suppose that element (I) will be not be used in a computation unless it is the largest element between index positions (I − 10) and (I + 10). Obviously, the processor cannot start by executing parallel, simultaneous computations on all of the elements of a 1 K vector. No matter how simple the rest of the algorithm is, the typical array processor cannot handle this type of job. The routine ends up as a conventional, scalar program.

Even if an array processor had the hardware to make such comparisions, such a procedure would be complicated by the fact that the pipeline in some small or medium-sized processors is only eight channels wide. The programmer is not aware of this limit, for there is hardware and firmware to repeat the operation in chunks, transparently, until the entire vector appears to have been handled as an entity. Internally, though, this limit would make it particularly cumbersome to do compare/select routines of greater span, as in the example given.

Nonuniformity of computations is another block to array processing. To illustrate, consider a vector that happens to meet the requirement for computational independence (i.e., where no calculation depends on the value of another element). But suppose the first seven elements are added to a constant, while the next 10 are multiplied, and a third group is raised to some exponent. In principle, of course, this could all be done in parallel, but no commercial AP is structured to handle such arbitrary mixtures in one pipeline stage.

Both properties—computational dependence and nonuniform mathematics—are common to vectors that represent devices with *memory*, especially if there is also nonlinearity. Examples include capacitor charge/discharge circuits, diode-capacitor circuits, devices with hysteresis, those with nonlinear feedback, adaptive and predictive circuits, and slew-limited devices. Such circuits are often used in measuring transient response, noisy signals, speech and music, and almost-periodic signals (those in which the components are individually periodic, but not coherent). For these, we are forced to consider algorithmic models that involve some degree of conventional, scalar, mathematics. Iterative, incremental algorithms provide a simple generic approach to modeling such circuits.

Incremental Modeling

Consider the simple RC charging circuit shown in Figure 11.1. The capacitor voltage rises at a rate $dV/dt = i/C$. In an incremental computer model, dt is represented by, or approximated by, a finite time interval: the time between two successive samples. For simplicity, we measure relative time by the number of increments; dt has a value of one.

The instantaneous current, i, is closely approximated by the average current that flows during the interval dt. For very small sampling intervals, however, the current will not change significantly during the time interval. In such cases, the computations are simplified by using either the initial or the final value of current for each interval, instead of the average. In the following examples, we will use the initial values.

Let the vector E represent the input voltage waveform and vector V the capacitor voltage waveform. E(I) and V(I) then represent the samples of input voltage and capacitor voltage taken at time I.

During the interval dt, the capacitor charge Q changes by the increment $dQ = i * dt$. Since dt is one unit, dQ has the same numerical value as i and, hence, as the expression, (E(I) − V(I))/R. Dividing this by capacitance C gives the incremental output voltage dV. The expression for dV is then

$$(E(I) - V(I))/T \qquad (11.2)$$

where T is the time constant ("Tau"), and equals $R * C$.

V(1) is the initial output voltage, here assumed to be zero. Successive samples of V may be computed iteratively, as shown here in BASIC:

```
V(1) = 0
FOR I = 1 TO N − 1
V(I + 1) = V(I) + (E(I) + V(I))/T          (11.3)
NEXT I
```

A plot of the response of this incremental model to a step input of 10 volts is shown in Figure 11.2. The step continues for four time constants. Each vector contains 512 samples. (All plots were made by an H-P 7470A X-Y Plotter via IEEE GPIB Bus.)

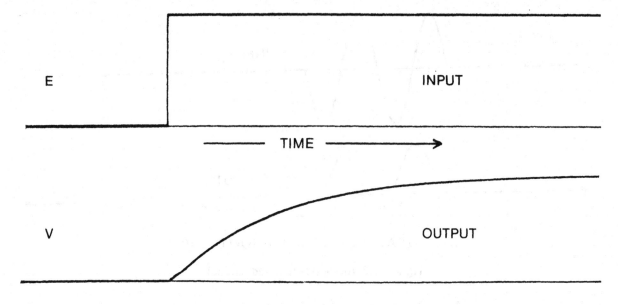

Figure 11.2: Response to Incremental R-C model to step input

The time constant used in Figure 11.2 is 96 increments; after 4T (= 384 increments), the voltage out of the incremental model has reached 9.8188. The voltage calculated by using the familiar exponential equation is 9.8168, an error of −0.02 percent.

Comparison with Continuous Equations

If all one needed to know was the voltage at a single instant, the classical exponential equation would be much faster, and more precise. The incremental model does not produce one value, however, but N values. It generates the sampled representation of a *function* (i.e., it creates a numerical vector). In producing an entire vector, the incremental model is faster than the exponential solution and requires only four-function (calculator-type) mathematics. It can thus be performed by inexpensive scalar coprocessing units. Ideally, such routines are built into the DSP software operating system and operate at machine level. They can also be implemented in real-time, outboard hardware.

An important advantage is that any input vector can be applied: complex waves, noise, transients, or periodic functions. Like a real resistor-capacitor (RC) circuit, it works in the time domain and produces a time-varying output wave. Moreover, the value of N is not critical, as with the FFT.

And its precision, while not perfect, is quite sufficient for simulating commercial resistor and capacitor networks.

Figure 11.3 shows the response of the above RC low-pass circuit to a gated sinusoid, 2 cycles at 1000 Hz. The time constant is 1.5 ms. This plot is an interesting one, a test of our understanding of linear networks. It is often said that a linear network (e.g., an RC circuit) cannot alter the shape of a sine wave, only the amplitude and phase. Yet the incremental model has obviously distorted the transmitted signal. Is the model wrong? Or is the theory wrong? Look carefully at the input before answering.

Neither is wrong, because the input is not a sine wave! A true sinusoid has no beginning or end; it is the only moving waveform whose shape cannot be distorted by an RC network. Such a wave is completely periodic and its spectrum is a single line. An RC network cannot add additional spectral lines.

Here, however, the input (at the top of Figure 11.3) is a *gated* sinusoid, having a complex spectrum. If the relative amplitudes and/or phases are changed, the waveshape will be altered. Although linear in voltage transfer, the RC network alters both the relative amplitude and the phase.

Here is another little test. You can see that the sinusoid itself is not continuous and periodic. But it is certainly pos-

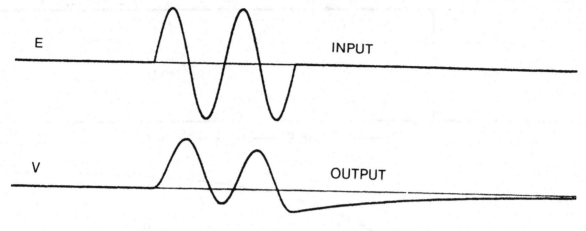

E   INPUT

V   OUTPUT

(OUTPUT AMPLIFIED X 4 TO SHOW DETAIL)

Figure 11.3: Response to gated sinusoid

sible for a discontinuous wave to be periodic; that is, the gated tone burst might be seen to repeat periodically if the plot extended far enough. If so, it would have a discrete (line) spectrum and could be analyzed by the coherent vector techniques described in the other chapters.

Under close examination, however, it becomes apparent that this burst is probably not periodically repeated, certainly not over an interval of time comparable to the plot duration. Do you see the clue?

The clue is in the filter response:

1. If the input to a linear filter (e.g., an RC network) is periodic, then the output will also be periodic, with the *same* period.

2. Since the right and left ends of the *output* response do not match exactly, the output is not periodic over the interval shown.

3. Since the output does not have that period, neither does the input.

4. Finally, since the filter response is zero right up to the gate *on* time, we know that there has been no recent input. Its repetition rate, if any, must be much longer than the gate interval. It is best to treat this gated sinusoid as a transient.

All electrical signals can be distorted in shape by classically nonlinear transfer functions. But a signal like the above—in fact, any wave that is modulated, gated, or composed of multiple tones—can also be distorted in shape by classically *linear* circuits if its phase-versus-frequency plot is not a straight line. Such circuits are said to have *nonlinear phase*, and the category includes familiar RC networks.

Whenever modulated waves are sent through them, they introduce *group delay or modulation distortion*. With AM or gated waves, the envelope distortion is readily seen, as in Figure 11.3.

To further check on the performance of our incremental model, however, note that a classically linear network cannot produce a DC component if there is none in the input signal, regardless of what it does to the envelope shape. Happily, neither does the incremental model. The area under the curve is nonzero (= 0.0050) only because the curve was not permitted to run its course. The computer data show that after about 12 time constants, the DC value falls below the resolution of the computer.

Such things may not be directly important in a particular test, but they show that the incremental models respond to transients just as RC circuits are supposed to. They are also mentioned to encourage you to perform similar checks whenever you use computer models. As soon as you install any model, test it by applying vectors with known characteristics before continuing to write the rest of the program.

In line with that advice, let's try one more test, applying a continuous sinusoid of 10 V amplitude (20 V p-p). Let the time constant equal half the period. By calculation, the amplitude output should be 3.033 volts (Figure 11.4).

This time, the model is not as precise as before, although still comparable to good capacitor tolerances. The output amplitude after 12 time constants (12 half-cycles, in this case) is 3.077—an error of 0.044 V from 3.033—almost 1.5 percent or 0.125 dB. This error, like the previous ones, derives

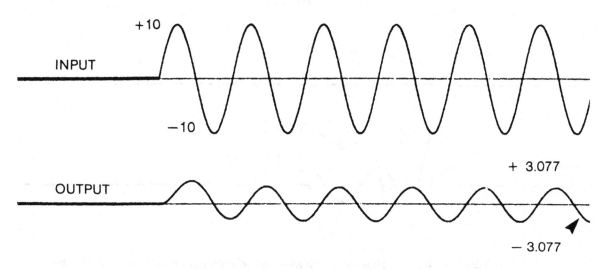

Figure 11.4: Response to sustained sinusoid

partly from the shortcut of using initial values instead of average. Rather than use the longer expression, however, one can often "tweak" the time constant slightly to bring the model closer to ideal.

RC High Pass Model

The previous FOR-NEXT loop can be used with vector subtraction to model a high-pass network (Figure 11.5). In the high pass network, the output signal vector V1 is the resistor voltage. To find this, subtract the capacitor waveform V from the input waveform:

$$\text{MAT V1} = E - V \qquad (11.4)$$

The response is plotted in Figure 11.6 for a step input (top) and for the gated sinusoid (bottom). The time constant was shortened so that the ratio of cutoff frequency to input frequency is the inverse of the low-pass ratio. This gives the same theoretical attenuation in both networks: a factor of 0.3033. Note that the error is essentially the same in both cases.

As an interesting aside, note that the input truly leads the output, exactly as theory predicts. Many engineering students, on first encountering lead networks—and knowing that nothing can emerge from a transmission circuit before it goes in—tend to believe that a leading output is really one that lags so much that it simply appears to lead a subsequent cycle! Not so. As you see, the peaks and zero crossings of the output lead from the very beginning. On the other hand, you will be gratified to see that nothing actually emerges from the output before the input is applied.

A final observation has to do with *group delay* (also called envelope delay or modulation delay). First, there is *delay dis-*

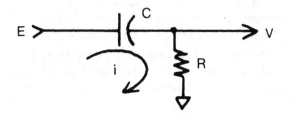

Figure 11.5: High pass RC network

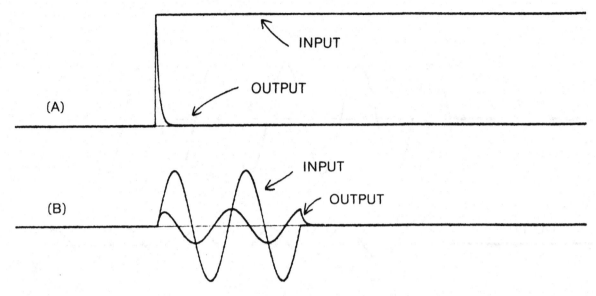

Figure 11.6: Input-output plots of high pass model with step input(top) and gated sinusoid (bottom)

tortion, whose symptoms include (in this example) a change in shape, most noticeable at the trailing edge. There is also *absolute* delay, a shift in the "center of gravity" of the output wave group, compared to that of the input. Even when the "carrier" sine wave has a leading phase, as it does here, the absolute delay to a transient must always be postive (i.e., a delay). This is in accordance with information theory. A "message," or informational package, must always experience a finite delay in passing through a real circuit—that is, one having finite dimensions and/or finite, nonzero electrical parameters. This delay is independent of the phase shift of the carrier component.

Time Scaling and Normalization

In modeling any network incrementally, it is convenient to scale all time values relative to dt. For example, in Figure 11.4, the period of the sinusoid is 1/8 of N = 512, or 64 increments. The time constant was stated to be 1/2 the period, so that in the actual program used to make the plot, T = 32.

In the gated sinusoid example in Figure 11.3, the signal period was 1.0 ms, and the time constant 1.5 ms. In this program, the value of T was therefore 1.5 × 64 time units, or 96.

Please note that T does not have to be an integer. In the high pass example, T is $32/\pi^2$, or 3.2423. With pulses and gates, it is generally good practice to make T at least five time increments, preferably 10 or more. This is done by increasing N as necessary.

Ballistic Peaks

In the days of D'Arsonval and moving iron meter movements, measurements were indicated by the position of the pointer, or "needle." Dynamic readings depended on what is called the *ballistic* behavior of the movement. The movement had mass and the springs had compliance. The mechanical resonant circuit that resulted was damped both magnetically and by nonlinear air resistance. According to the source resistance seen by the coil as it looked back into the divider, driver, or rectifier circuits, each movement had a characteristic rise time, overshoot, and settling. The release or fall was not necessarily symmetrical with the rise but was often slower.

This worked out well for reading sound level peaks, because the meter, like the ear, would not respond fully to very short peaks. In addition, inertia and damping of the needle would hold it near the indicated value long enough for the eye to make a useful reading.

If we plotted a graph of overall impulse response, the result would resemble the path of a bullet in or cannonball through the air, a so-called *ballistic* trajectory. As a consequence, the dynamic behavior of the movement was often referred to as its ballistic behavior.

It is this behavior, for example, that defines the difference between a dB meter and a "volume unit" (VU) meter. Both are supposed to be true RMS-responding meters and, with steady state waveforms, give identical readings. But the dB meter is specified *only* for steady levels, whereas the VU meter, upon application of a step wave, must rise to 99 percent of its ultimate value in 0.3 seconds, and exhibit 1.0 to 1.5 percent overshoot. It must give a reading of 80 percent on an impulse of sine waveform (a gated sine wave) of 25 milliseconds duration, with sinusoids from 35 Hz to at least 10000 Hz! This ensures that all VU meters respond dynamically the same way to program material: speech and music.

These numbers are not important to the immediate problem, but they represent the kind of parameters you will have to consider—and be able to model—if you are called upon to use DSP to replace certain analog instruments. If the device specifications call for any measurement in VU, you would need to use the ballistic time weighting parameters, and know how to incorporate them into a suitable mathematical model. In performing wow and flutter measurement, the peak-reading portions of a DIN/IEEE standard flutter peak detector must also conform to ballistic specification, one we shall examine later. In any case, ballistic behavior means that the calculations performed on each sample are dependent on previous values and thus are candidates for incremental algorithms.

Ballistic Peak Detection Models

To simulate a delayed charging circuit, let us add a diode to a low pass RC circuit (Figure 11.7). This circuit can charge, but not discharge. To a short pulse, it will charge only part way. In the following BASIC routine, D is the forward diode drop.

```
10   V(1) = 0
20   T = 40
30   FOR I = 1 TO N − 1
40   IF E(I) > (V(I) + D) THEN GO TO 60        (11.5)
50   V(I + 1) = V(I)
60   V(I + 1) = V(I) + (E(I) − V(I))/T
70   NEXT I
```

Figure 11.8 plots the response of this charging circuit to a pulse of width T. For simplicity, the diode drop D has been set to zero. This routine produced a final voltage of 6.3677 volts with a 10 volt pulse input. A perfect model would produce 6.3212 volts.

To approximate ballistic behavior, we can provide a discharge path for the capacitor by way of a second resistor R2 (Figure 11.9). The charging path is through R1 alone, while the discharge occurs through R1 + R2. In this particular arrangement, the discharge time is greater than the charge time, which is typical of many ballistic measurement specifications. The ratio of charge to discharge time constants is

$$\frac{T1}{T2} = \frac{R1}{R1 + R2} \qquad (11.6)$$

To simulate this behavior, line 50 of the previous BASIC routine is modified to allow a slow discharge of the capacitor D = 0.

```
10   V(1) = 0
20   T = 40
30   FOR I = 1 TO N − 1
40   IF E(I) > V(I) THEN GO TO 60        (11.7)
50   V(I + 1) = V(I) − V(I)/T2
60   V(I + 1) = V(I) + (E(I) − V(I))/ T1
70    NEXT I
```

The reader should recall that there are similar algorithms based on average rather than initial values. But we must be careful in when time (I) and time (I + 1) both appear on the right side of the equations, because there are many innocent-looking iterations with this property that actually oscillate or latch with certain input patterns. Initial-value iterations are generally the safest to use.

Figure 11.10 shows the plot of vector V where the input E is a pulse of width T1. The discharge constant T2 is 5 * T1.

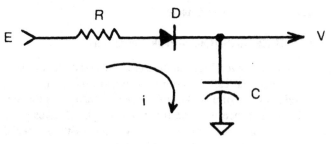

Figure 11.7: Charge-hold circuit

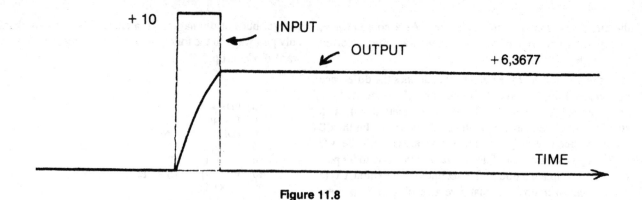

Figure 11.8

Generalized Approach

The rest of this chapter deals specifically with circuits for "wow" and "flutter" testing and is of limited interest. For the reader who wishes to stop at this point, the previous examples should suffice to illustrate the fundamental approach to incremental modeling and to suggest some of the checks we should apply in debugging them.

To recap, incremental models apply chiefly to circuits that have memory (e.g., analog-digital functions with counters or RAM functions), and to those that store energy (e.g., R-C and R-L circuits). Any analog function that can be expressed by a differential equation is a candidate for incremental simulation. Time responses that involve overshoot or ringing can be approximated by simple damped resonance (R-L-C), and thus be incorporated into these models.

Wow and Flutter Measurement

"Wow" refers to unwanted speed or frequency modulation at rates below about 6 Hz. "Wow" is a good example of onomatopoeia. The term was introduced in the 1930s to describe (by its American or English pronunciation) the sound produced when playing a conventional disc recording in which the hole had been punched off-center. A rim-drive turntable can produce the same effect if the rim is not perfectly centered on its axis. At 33-1/3 RPM, the wow rate is 0.555 Hz.

The slowest wow-producing element in a cassette tape deck (transport mechanism) is usually the pinch roller, producing modulation in the 1 to 2 Hz region. Lower frequency modulation would most likely result from a faulty cassette mechanism and so would probably not be part of the transport performance specification.

"Flutter" refers to FM at rates from 6 Hz upward. In some texts, this range is subdivided into a primary zone from 6 to 30 Hz and into a secondary zone from 30 to 100 Hz. This is because modulation by components that could be audible by themselves have a different subjective effect.

In tape decks, flutter in the 6 to 30 Hz zone can be caused by an out-of-round capstan or capstan pulley. Variations in the higher zone are most likely to be related to motor vibration or an out-of-round motor pulley. If the motor speed were 1800 RPM, for example, an out-of-round pulley would modulate tape speed at a rate of 30 Hz. Frequencies around 40 to 50 Hz (2400 to 3000 RPM) are commonplace for the DC motors used in automotive cassette players.

The wow and flutter spectrum of a typical automotive tape deck was plotted in Chapter 7.

DIN/IEEE/ANSI/Quasi-Peak Detection

There are several related measurement standards for wow and flutter that call for both *frequency-weighted* and *time-weighted* (ballistic) peak measurements. Insofar as the signal flow is concerned, the frequency weighting takes place first, but for tutorial purposes, we will start with the special ballistic, or *quasi-peak*, detection circuit.

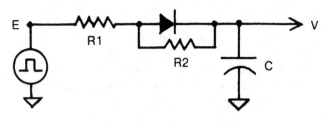

Figure 11.9: Ballistic circuit

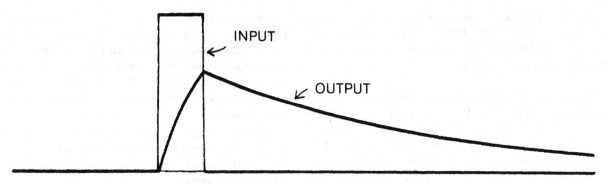

Figure 11.10: Ballistic response to pulse input

In DIN (Deutsche Industrienorm) quasi-peak detection, the frequency modulation (FM) detector output is spectrally weighted to match the ear's sensitivity and then is sent to two parallel ballistic peak detectors: one for positive voltage (i.e., positive frequency deviations) and the other, with its diode reversed, for negative voltage. The two ballistic waveforms are combined by a difference amplifier to provide the sum of the absolute voltages (Figure 11.11).

If there were no discharge paths, the output would eventually climb to the peak-to-peak voltage. Because of the continual discharge, however, and because peaks of opposite polarity cannot occur simultaneously, the output never reaches that value. *A quasi-peak detector provides a histogram of short-term frequency deviation.*

The previous BASIC routines may be expanded to simulate this action. In this example, vector A represents the positive capacitor voltage waveform, and B is the negative. The time constants are approximately 30 ms for charging and one second for discharging. At 315 samples per second, the rate often used in measuring tape FM by event digitizing, one time unit = 3.17 ms, so T1 = 9.5 and T2 = 315.

```
30    FOR I = 1 TO N − 1
40    IF A(I) < E(I) THEN GO TO 60
50    A(I + 1) = A(I) −A(I)/T2
60    A(I + 1) = A(I) + (E(I) −A(I))/T1
70    IF B(I) > E(I) THEN GO TO 90        (11.8)
80    B (I + 1) = B(I) − B(I)/T2
90    B(I + 1) = B(I) + (E(I) −B(I))/T1
100   NEXT I
110   MAT V = A −B
```

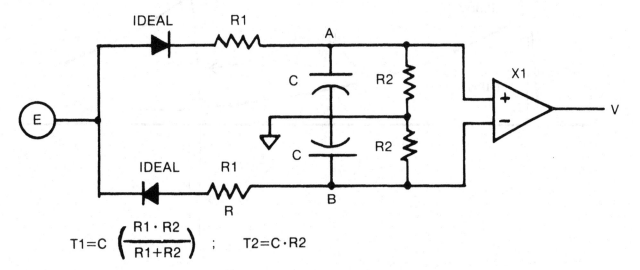

$$T1 = C\left(\frac{R1 \cdot R2}{R1 + R2}\right) \quad ; \quad T2 = C \cdot R2$$

Figure 11.11: Bipolar ballistic (quasi-peak) detection

Note that the final operation, line 110, is a *vector* operation. Unlike the RC-diode circuits, the difference amp is a time-invariant (memoryless) circuit and can thus be simulated by a vector subtraction.

The quasi-peak response to pulses of $+10$ and -10 volts is plotted in Figure 11.12. Pulse width is 17 units, slightly less than $2 * T1$. The first pulse charges capacitor A to 8.96 volts, and the second pulse charges capacitor B to -8.96 volts. Note that a true peak-to-peak detector would hold these values and add their magnitudes, giving 17.92 volts. The maximum quasi-peak output is only 10.16 volts, however, because the two pulses occur at different times. (A) falls to 1.20 volts by the time (B) reaches maximum charge; the sum is $1.20 + 8.96 = 10.16$.

In the complete instrument, the quasi-peak output is followed by a true peak-hold circuit that captures the *highest quasi-peak during the measurement interval*. This is the DIN/ANSI/AES-defined flutter. In the DSP equivalent, it is determined by the operation,

$$\text{MAT P} = \text{XTRM (Q)}. \qquad (11.9)$$

For purposes of statistical analysis, flutter meters from manufacturers such as Rohde and Schwartz also include a "two-sigma" program, in which the quasi-peak output is sampled at 100 ms intervals for 5 seconds. The three highest values are discarded, and the fourth largest is recorded. For normal (Gaussian) distributions, this value is twice the standard deviation.

In a DSP system, the same thing can be done by restricting the range of the XTRM operation to a zone of approximately 100 ms and then by repeating the XTRM operation for different zones.

Since the mean of MAT Q is zero, however, the standard deviation can be more quickly found by the operation, MAT R = RMS(Q). This value is also correct for non-Gaussian distributions, which is not the case for the above scheme.

DIN Frequency Weighting

The perceived intensity of wow and flutter depends on the frequency, reaching a maximum around 4 Hz. The German DIN standards require that the sensitivity of the quasi-peak detector be made to vary accordingly. This curve is essen-

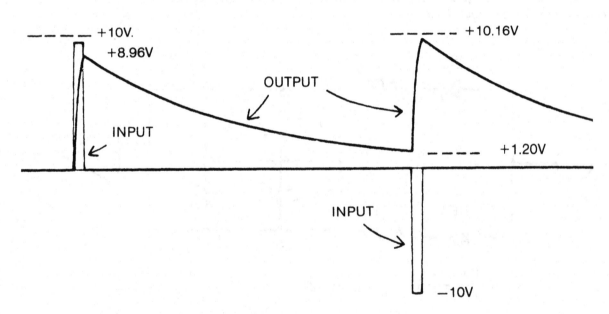

Figure 11.12: Response of quasi-peak detector to bipolar pulse input

tially the same as prescribed by the European *CCIR Recommendation 468, IEC Pub. 386*, the American *ANSI S4.3-1982*, the *AES 6-1982*, and the *IEEE Std-193*.

The recommended frequency-weighting curve is shown in Figure 11.13. A weighting curve that fits within these limits can be produced by a low-pass RC filter section in cascade with two identical high-pass sections. The three dB frequencies are 12, 0.95, and 0.95 Hz. (Figure 11.14).

If desired, this model can be implemented by three incremental computing loops. The elementary loop is a low-pass filter whose input is MAT E and whose output is MAT V. The high-pass section is the same loop followed by the vector subtraction of V from E. In the following procedure, the output from each section is relabeled E to serve as the input of the subsequent algorithm.

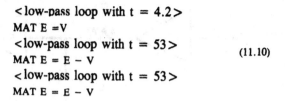

```
<low-pass loop with t = 4.2>
MAT E = V
<low-pass loop with t = 53>
MAT E = E − V
<low-pass loop with t = 53>
MAT E = E − V
```
(11.10)

Importance of Phase Response in Peak-Reading Instruments

The incremental model has the advantage that it treats real signals—those that may be poorly periodic or random—in the same way as real analog hardware filters. Its disadvantage is that it is somewhat time consuming as a high-level routine.

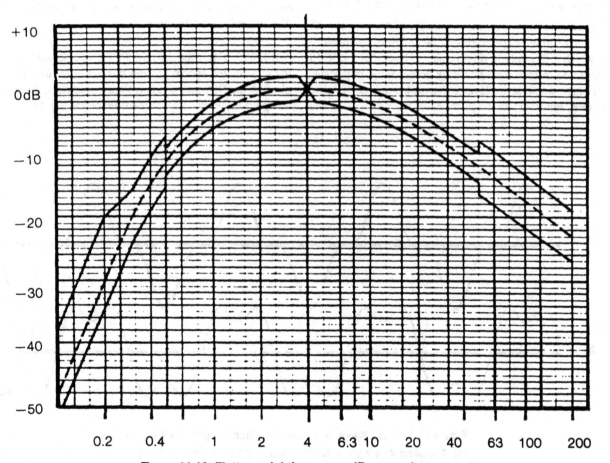

Figure 11.13: Flutter weighting curve: dB versus frequency (Hz)

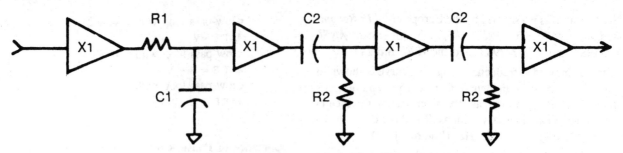

Figure 11.14: RC approximation of DIN 45 507 response

In Chapter 7, the FFT was used to provide the DIN/AES frequency weighting. This is a rapid procedure and was acceptable there because the main parameter of interest was RMS flutter. Where the intent is to find the quasi-peak response, though, that simple procedure is insufficient. Unless the *phase* and *impulse response* characteristics of the DSP filter match those of the filters in the analog standards, the time waveshape of the weighted flutter error will be different, and the quasi-peak readings will be different.

Figure 11.15 shows a simple example of two waves with the same magnitude spectrum—a first and third harmonic—and which differ only in relative phase. Note that the RMS is the same in both waves; if only power were important, filter phase would be irrelevant. The peak amplitude is quite different, however; so is the *shape* of the peaks (i.e., the

amplitude distribution), and that further affects the output of a ballistic peak detector.

Filter phase is not specified in any of the formal standards, but those standards assume active or passive analog filter designs. Most of these behave very much in phase and impulse response like the RC model in Figure 11.14. The usefulness of this model is not necessarily as part of the test program itself, but as a means of determining the phase and/or impulse response that should be assigned to the DSP filter actually used in the program.

If flutter were coherent, the FFT could be used to obtain the unweighted spectrum, and this could in turn be multiplied by the DIN magnitude response, provided the spectral components *were also shifted in phase* by the amount

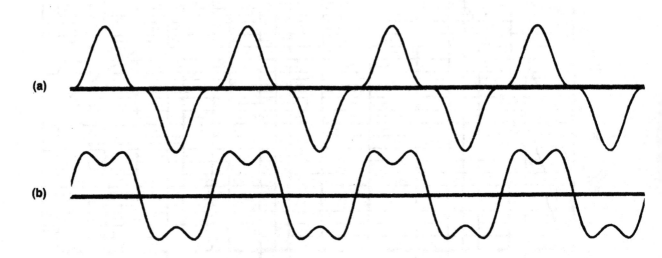

Figure 11.15. Two Waveforms with Identical Magnitude Spectra
(a) Absolute Average = 1.698; RMS = 2.236; Pk-Pk = 8.000
(b) Absolute Average = 2.122; RMS = 2.236; Pk-Pk = 5.756

expected in a conventional analog, RC filter. Because flutter is not coherent, however, the FFT by itself does not provide an accurate spectral picture even with proper phase.

In Chapter 7, a time window (Hann) was applied to the FM signal before using the FFT. This modulates the signal by compressing its amplitude to zero at the ends of the window, making the looped pattern smoothly periodic over the FFT period. Windowing makes the resulting spectrum a good approximation of the true spectrum in *magnitude* distribution and in total power. Unfortunately, a time window *alters the amplitude distribution of a waveform, and thus alters the peak-to RMS ratio of complex waveforms.* As a consequence, the window-FFT approach may not be suitable for frequency weighting in *peak* measurements, or *ballistic* measurements.

Finite Impulse Response (FIR) Filtering

A useful machine-level alternative available in many DSP operating systems is the FIR filter. This is also an incremental time-domain operation, although one based on *impulse response* instead of incremental charge-discharge. But the fact that it is a time-domain operation does makes it a useful filtering tool for transients and quasi-peak measurements.

FIR operation is explained in references [7], [8], and [9]. Essentially, it treats any sampled waveform as a sequence of *impulses* whose areas are proportional to the sampled voltage. If the filter's response is first established for a single impulse, then the output to a sampled waveform can be represented as a series of overlapping (superimposed) impulse responses, each of which is scaled (multiplied) by the sample amplitude.

Although somewhat slow because of its sequential nature, FIR filtering is a machine-level routine performed by many array processors. Here is one syntax example:

$$\text{MAT } Y = X \text{ FILTERED WITH } C \qquad (11.11)$$

Vector X is the input signal and vector Y is the filtered output waveform. C is the list of coefficients, which is nothing more than the sampled waveform produced by the the application of a *single impulse*. This may be obtained directly from the incremental charge model by applying an impulse vector: a "one" followed by $N-1$ zeros.

$$\text{MAT } E = [0]$$
$$E(1) = 1 \qquad (11.12)$$

The details of applying FIR filters go beyond the scope of this article. We should note, however, that there are often two forms available: an *open*, or transient form, and a *closed*, or periodic form. In the first, the output dimension is greater than that of the input vector, reproducing the stretching of transients characteristic to *all* real filters, regardless of type. In the closed form, the output vector is looped back and superimposed on itself, so that the dimension is shortened to *match the period of the input*. With linear filters, periodic inputs produce identical output periods. In flutter testing, it is the *open* or transient form that is needed.

Another comment is that the FIR algorithm is sequential and therefore slow when compared with operations like the FFT. For typical N, the FIR model may take longer to compute than the incremental model if MAT C has a comparable number of points. To reduce computing time, it is worthwhile to experiment with short vectors for C, which you can obtain by artificially rolling off the tail of the impulse response. By adding enough zeros so that DIM(C) is a large power of two, you can use the magnitude FFT to compare the resulting frequency response with the DIN tolerance template. With a little smoothing, it is possible to get sufficiently accurate DIN filtering with less than 50 samples.

Chapter 12
CODEC Testing

Chapter 12:   CODEC Testing

We would expect a mass-produced, voice frequency (VF) component to be easy to test. But in the case of the CODEC, it is quite the opposite. Codecs encompass a wide range of physical principles: linear amplification, sampling, quantization, nonlinear encoding, waveform reconstruction, active filters, and logic functions, to name a few. In addition, their performance standards are often alien to those who manufacture or test them. To automatic test equipment (ATE) makers and users alike, codecs pose a complex and intriguing challenge.

Digital signal processing (DSP) measurement techniques for voice frequency (VF) parameters—gain, signal-to-distortion, harmonic distortion, intermodulation (IM) distortion, random noise—have been previously shown. In this chapter, the emphasis is on the terminology and measurement problems peculiar to digital VF telecommunications measurements, especially those involving mu-law and A-law encoding.

The international codec performance standards were established at a time when measurements were made by analog instruments. To emphasize the intent of these standards, some of the test diagrams are shown in their conceptual, or analog, form. With DSP-based test systems, the analog filter and voltmeter function will be simulated by software models, as will reference encoders and decoders.

The Pulse Code Modulation (PCM)/CODEC Channel

The term *codec* denotes the related pair of digital *coding* and *decoding* functions in a PCM telephone channel (Figure 12.1).

Two codec pairs are required to complete a bidirectional channel. Each pair permits analog voice and tone signals to be multiplexed, switched, and transmitted in digitally encoded form and greatly simplifies the requirements for these functions.

From a transmission viewpoint, we may define a codec as the pair in either of the two horizontal "channels." *For test and manufacturing purposes, though, the codecs are considered to be the pairs at either end.* The significance is that the pair we can test is *not* the pair that will communicate

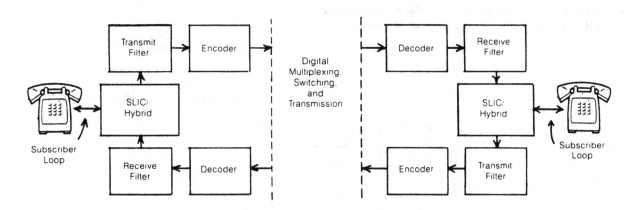

Figure 12.1: PCM telephone channel

as a channel, a fact that is all too often forgotten in codec testing.

In a digital telephone system, the telephone itself remains an analog component. It is connected via a bidirectional *subscriber loop* to equipment at a *central office* (CO) or *private branch exchange* (PBX). Here the path is separated into individual *transmit* and *receive* paths by a 2 to 4 wire "hybrid," which is part of the *subscriber loop interface circuit* (SLIC).

After separation, the signal to be transmitted is first band limited to 300 to 3400 Hz, removing power line and ringing frequencies, and eliminating most of the signal energy at 4000 Hz and above. This permits the encoder to sample the filtered signal at 8000 samples per second, the worldwide standard. Each sample is then converted to an 8-bit digital "word" and transmitted serially and multiplexed with words from other PCM channels.

At 8000 words per second, each single-channel encoder has an *average* data rate of 64 kilobits per second (kb/s. In a complete system, however, each sample is transmitted as an 8-bit burst at a much higher momentary rate (e.g., 1544 or 2048 kb/s). This is done to allow time multiplexing of 24 to 32 channels into a so-called *primary* group, which may be directly transmitted for short-haul applications or be further multiplexed for longer-haul use.

The receive side must unravel the foregoing process, first by selecting the burst from the proper time slot, then by decoding each burst into one of a set of discrete voltage levels. The resultant stepwise waveform is fed to the *PCM receive filter*, which reconstructs a smoothed, continuous approximation to the original audio signal (Figure 12.2).

The receive filter also applies a correction for amplitude and phase to compensate for the "sine-x-over-x" frequency rolloff of the stepwise waveform. Done properly, the effects of sampling time quantization are removed, and the only significant errors are those of amplitude quantization.

Encoding Law

To complicate matters, the 8-bit coding is not a simple linear conversion but a quasi-logarithmic one that compresses the signal range according to the "mu-law" (North America, Japan, and related systems) or the "A-law" (Europe and related systems). Figure 12.3 shows the encoding curve for mu-law).

Each quadrant of the mu-law (μ-law) is a piecewise approximation to a semi-logarithmic function and is composed of eight *segments*, or *chords*. Within each chord are 16 uniform steps. Moving away from the origin, each chord has twice the width of the preceding one, so the steps themselves double in width every 16 steps. In A-law, the first two chords are uniform but are doubled thereafter as with mu-law.

The intent of either "law" is to provide quantization roughly proportional to the signal size. This form of instantaneous signal compression results in a fairly uniform signal-to-quantizing noise ratio, roughly 40 dB over a wide range of input amplitudes. Because the resolution improves with small signals, the dynamic range is quite good: nearly 80 dB, or that of a 13-bit linear A/D, for mu-law. We may thus say that mu-encoding compresses a 13-bit dynamic range into an 8-bit format.

The decoder performs the inverse function, *expansion*. Discounting the quantization, this results in a linear composite transfer function for the transmission pair. The combined action is termed *companding*. Any mismatch between the two curves shows up either as nonlinear distortion or as gain variation with changing input level.

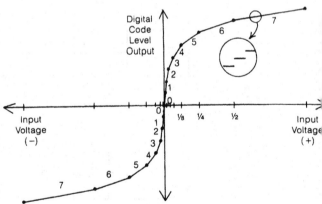

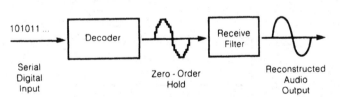

Figure 12.2: Reconstruction of digitized signal

Figure 12.3: Mu-law encoder transfer function

In mu-law, the CODEC pair must also permit *signaling*, or the transmission of loop status and similar low-frequency information. For this, the code format is altered every sixth frame, substituting a signaling bit for the normal least significant bit (LSB). Not only does this give another logic function to be tested, but it also requires a specified analog shift in the A/D decision levels and D/A output levels. This alters the audio performance and adds additional testing in this area too. Signaling is not simply a logic function.

Five Kinds of Tests

There are at least five categories of tests that must be provided in a complete CODEC test system: performance test, digital functionality, DC and AC parametrics, high-speed margin testing, and characterization tests.

1: PCM Performance Tests

Performance tests evaluate each device under test (DUT) strictly in terms of its projected effect on the transmission of voice or data signals through a complete channel. They are alternately called *transmission* tests. Examples include gain, signal-to-noise, nonlinear distortion, and group delay distortion.

Although routine to other kinds of components, such tests are far more difficult in the codec case. A surprising amount of accuracy is required and yet production volume dictates that such parameters be measured in milliseconds. Remember, too, that the channel is not all analog. There are many interactive effects of sampling and quantization that interfere with each reading. Finally, these audio tests must be applied to individual sections of a channel, including the encoder and decoder (i.e., devices that are part analog and part digital).

The performance requirements for in-circuit codecs are established internationally by the International Telegraph and Telephone Consultative Committee (CCITT), and within each country by the appropriate administration (e.g., AT&T (USA), Bell Canada (Canada), etc.). At the component level, the manufacturer will apply tighter limits as part of an overall error budget.

2: Digital Functionality Tests

Whereas performance tests measure the parameters of voice and data transmission, digital functionality tests simply ensure that all of the logic functions built into particular component are working. Such tests include signaling functions, power down, time slot assignment (if present), A-law

or mu-law selection, test mode selection, and so on. The answers are strictly "pass/fail." To provide these functions a test system needs programmable digital test capability, plus the necessary "personality" interface circuitry.

3: DC and AC Parametrics

Parametrics are the functions we normally associate with conventional linear ATE systems: shorts, opens, leakage, logic levels, thresholds, ripple rejection, impedance, supply sensitivity, and so on. So called parametric testing usually involves only one "port" of the circuit, in contrast to performance or transmission testing, which involves two ports (input and output). The answers are in traditional electrical units like volts, kilohms, and microamperes. Although parametrics are conceptually simple, the total test circuitry required may exceed that for DSP-based transmission test. Bench transmission-test instruments do not usually provide parametric testing.

4: High Speed Parametrics and Margin Tests

There are two goals in high-speed testing. One is to measure timing parameters called out on the specifications sheet, such as rise time or propagation delay, while the device is operated with nominal timing. The other is to vary, under-program control, the digital stimuli, and to find out how much leeway the device provides.

The latter procedure is called margin testing and is a worthwhile form of insurance. It extends the standard codec tests to show how close a good part is to failure or to identify failure mechanisms that may not be detected in a routine test. For example, single-chip codecs are often tested with the same clock for both transmit and receive sections. In a real system, however, there are two different clocks of drifting phase, and excessive on-chip coupling may cause malfunction at some critical phase. Margin testing can simulate this condition and, moreover, pinpoint the hazardous phase.

High-speed margin testing requires the functions normally associated with a large digital test system: individually programmable edge delays, clock frequencies, phases, comparator strobe width, and timing. The test head will structurally resemble that for a digital system, too, because the CODEC is invariably a low power, high impedance device that cannot directly drive transmission lines.

5: Characterization

The last test category consists of those standard tests that may be performed reliably on a sample basis, plus specialized tests that yield information required only for design anal-

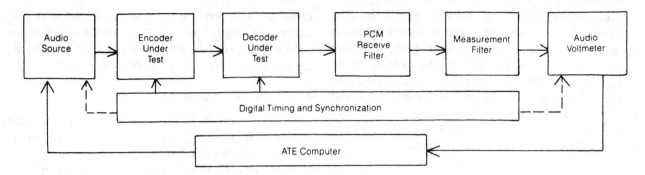

Figure 12.4: Typical full channel test fixture (analog concept)

ysis, process feedback, or vendor evaluation. Throughput is less important here. The emphasis is on accuracy and versatility.

A DSP-based test system has a particular advantage in characterization, for the test engineer may employ error correction techniques for extreme accuracy, and may use software to simulate a wide variety of test conditions not obtainable with standard hardware.

Full Channel versus Half Channel

Codecs were originally tested on existing analog transmission-type telephone instruments, adapted by the addition of digital clocking and synchronization. A test performed this way is an analog-to-analog test and must be made on a complete channel: encoder, decoder, plus filters. Such a test is termed a *full-channel* test (Figure 12.4). If the encoder and decoder are the common-end pair, the test is also referred to as a *self-loop* or *loop-around test*.

Randomly paired full-channel testing is an important part of testing a complete VF PCM switching system, since the pairing of transmit and receive halves in a real system is arbitrary and changes continually.

The only pair that will never communicate is the *self* pair. For this reason, loop-around or self-loop testing is inadequate as a codec test. It is necessary to test each channel half *alone* in order to identify its individual error contribution.

Half-channel testing is particularly important with commercial codec designs because the self pair often has *complementary* errors; the pair may pass all tests, yet each half may be out of specification. The direction of symmetry is random from unit to unit, however, so self cancellation of errors cannot be counted upon in real (random-paired) switching systems.

In summary, a good codec test system must not only provide the five categories of testing but also must be able to

implement them for either the full-channel or the half-channel case.

Normalized Mu-Law and A-Law Measurement Units

Each of the many subsections in a CODEC system is allowed to operate at its own optimum voltage or current level. But all units will be calibrated by outboard gain or attenuation so they share a common point of reference: the *virtual edge*, or formal limit to the signal range covered by the encoding law.

Prior to encoding, there is a continuum of instantaneous signal values within the signal range. For uniformity, these values are measured in so-called *normalized* mu-law or A-law units, chosen so that all significant points fall at integers and so that the smallest difference of such points is one unit.

Figure 12.5 shows the composite transfer curve of a mu-law encoder/decoder pair around the origin. It plots discrete output voltage versus continuous input voltage. The A/D decision levels occur at odd-numbered positions when meas-

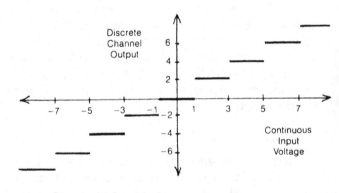

Figure 12.5: Mu-law channel transfer function in normalized units

ured in normalized units: in chord zero, at ± 1, ± 3 units, ± 5 units, and so on. Any input sample that falls between one and three units, for example, will be encoded as chord zero, step one. A voltage sample of magnitude 3.2 units will be encoded as chord zero, step two. This information, including sign, is transmitted as an 8-bit code word.

Upon receiving any code word, the decoder makes the best assumption that it can: that the original sample lay halfway between the two decision levels. In chord zero, the decoder thus produces the discrete analog output set, 0, ± 2, ± 4, ± 6, etc. Just as with the linear converters of Chapter 9, codec ADCs deal with code *edges*, while codec DACs deal with code centers.

Mu-law quantization differs from linear quantization, however, in that the step size (i.e., the difference between two A/D decision levels) doubles with each successive chord. The general formula for the *larger* of the pair of edges is

$$2^c * (2S + 34) - 33 \qquad (12.1)$$

where C is the chord number 0 to 7, and S is the step number 0 to 15.

Substituting C = 7 and S = 15, we obtain a level of 8159 units. There is actually no decision level here, because we have run out of code permutations. But 8159 does define the end of the last step: It is the *virtual edge*.

In the mu-law system, a sinusoid *at the codec* whose peak value is 8159 units, corresponds to a sinusoid *at the center of the central office switch* whose power level is 3.17 dB above 1 milliwatt. In other words, the maximum signal, 8159 units, has a *relative* level of +3.17 dBm0 (not dBm, which is absolute). We may thus relate all lesser signals to either normalized units or dBm0, as is appropriate to the test.

The European A-law differs somewhat from the North American mu-law. For the first chord of A-law (C = 0), the decision levels are 2 * (S + 1). The formula for chords 1 to 7 is $2^c * (S + 17)$. The virtual edge is 4096, and the corresponding power level is +3.14 dBm0 in A-law.

Review of Decibel-Based Measurement Units

Nearly all VF performance parameters are expressed as decibel ratios (dB), using a reference denoted by a suffix (e.g., dBm0). These references are not necessarily those used in other areas of engineering, so it is best to review them before continuing.

The Decibel

The dB is a unit of ratio, and for computation requires two arguments, or input values. If N represents the number of dB, then for power,

$$N = 10 \log \left(\frac{P_2}{P_1} \right) \qquad (12.2)$$

Most CODEC signal measurements are made in voltage. Since this is related to the square root of power when the impedance is constant, we may write

$$N = 20 \log \left(\frac{V_2}{V_1} \right) \qquad (12.3)$$

Step	Chord 0	1	2	3	4	5	6	7	Step
0	1	35	103	239	511	1055	2143	4319	0
1	3	39	111	255	543	1119	2271	4575	1
2	5	43	119	271	575	1183	2399	4831	2
3	7	47	127	287	607	1247	2527	5087	3
4	9	51	135	303	639	1311	2655	5343	4
5	11	55	143	319	671	1375	2783	5599	5
6	13	59	151	335	703	1439	2911	5855	6
7	15	63	159	351	735	1503	3039	6111	7
8	17	67	167	367	767	1567	3167	6367	8
9	19	71	175	383	799	1631	3295	6623	9
10	21	75	183	399	831	1695	3423	6879	10
11	23	79	191	415	863	1759	3551	7135	11
12	25	83	199	431	895	1823	3679	7391	12
13	27	87	207	447	927	1887	3807	7647	13
14	29	91	215	463	959	1951	3935	7903	14
15	31	95	223	479	991	2015	4063	8159	15

Figure 12.6: Mu-law A/D decision levels. The outermost level of each step is shown. Decode levels are midway between decision levels.

Relative Level

Level measurement requires only one argument. For normal signals, level is measured in dB with respect to 1 milliwatt, using the mnemonic "dBm." In telephone systems, however, it is not the power at the point of actual measurement that is meaningful, but the power at the center of the central office "switch." The point is called the *zero transmission level point* (0TLP). Anything we measure at the codec must be translated to the 0TLP. The unit of *translated* level is the dBm0. (The suffix is pronounced "oh.")

To illustrate, suppose we are told that a signal of 1.5 V root-mean-square (RMS) at point X will cause a level of zero dBm at the 0TLP. If we measure 2.0 V RMS at point X, then the *relative* level, expressed in dBm0, is

$$L_x = 20 \log (2.0/1.5)$$
$$L_x = 20 (1.333) \qquad (12.4)$$
$$L_x = 2.5 \text{ dBm0}$$

We do not need to know the impedance at X or the power. We need only the initial information relating V_x to 0 dBm0. As long as we stay at X, the value "1.5" is constant. Other points have other references, of course.

Note that there is no standard voltage for the codec, *nor should there be one.* Each type of codec is allowed to operate at a level best suited to its semiconductor technology. It is easy enough for the switch designer to apply the proper gain or loss exterior to the codec.

Although there is no absolute voltage standard, there is a well-defined standard *relative to full scale.* The voltage at which the codec begins to clip is called the *virtual edge.* It is measured in relative or *normalized* voltage units, chosen so all quantization levels are expressed by integers. For mu-law, the edges are ± 8159 units. For A-law, the edges per CCITT are ± 4096 units.

A sinusoid whose peaks touch the mu-law edges has a relative power level of $+3.172$ dBm0. The corresponding A-law sinusoidal level is $+3.14$ dBm0. These numbers are chosen to define full-scale sinusoids because they minimize the intrinsic gain error for tests at *zero* dBm0 (only) at 1000 Hz.

On the normalized scale, a sine wave at zero dBm0 has an RMS (not peak) value of 4004.3 mu-law units, or 2017.7 A-law units. In these universal units, any complex waveform whose RMS value is U units, has a relative power level of approximately

$$L_\mu = 20 \log (U/4004) \text{ in dBm0} \qquad (12.5)$$

or

$$L_A = 20 \log (U/2018) \text{ in dBm0} \qquad (12.6)$$

When the level is computed by DFT, FVM, or FFT, the *amplitude* (peak value) is obtained, not the RMS. For peak sinusoidal voltages, the normalized references for 0 dBm0 are 5662.9 (mu-law) and 2853.4 (A-law). The dBm0 level for a sinusoid of U(peak) normalized units is thus

$$L = 20 \log (U \text{ (peak)}/5663) \qquad (12.7)$$

Noise and Distortion

Noise components are measured with either *C-message* or *psophometric* filters. A level measured with either filter is denoted by adding the C or p suffix: dBm0C or dBm0p.

In North American systems, noise is frequently measured against a *picowatt* reference noise (rn), instead of 1 milliwatt. The unit becomes dBrnC0. There is a difference of 90 dB.

$$0 \text{ dBm0C} = +90 \text{ dBrnC0} \qquad (12.8)$$

dB Difference

Note finally that the difference between two levels expressed in like units is given in dB, *not the unit of level.* Thus the difference between $+13$ dBrnC0 and $+19$ dBrnC0 is 6 dB.

CODEC Performance Tests

Performance tests measure the *transmission* parmeters that affect both voice frequency (VF) and data signals in a telephone channel. These tests are based on classical audio measurement techniques, and their results are expressed either in dB, relative power level (dBm0), or subjectively weighted noise level (dBrnC0 and dBm0p).

In codec testing, transmission performance tests are applied to three circuit configurations: transmit half-channel (analog-to-digital), receive half channel (digital-to-analog), and full channel (analog-to-analog). The half-channel cases are generally the more important and difficult ones, but the full-channel case serves best to illustrate the concepts.

Gain and Loss

In telephone terminology, *loss* and *gain* measure the same property, namely, output-to-input ratio. In dB, they differ only in sign. The term, loss, is historically the correct unit, originating around 1900 when power gain (amplification) did not exist.

Ideally, no telephone connection should cause any overall gain or loss, regardless of length or type. To approach this ideal within reasonable bounds, each of the many elements that make up an arbitrary path or "connection" must exhibit a well-controlled gain, often to ± 0.1 dB or better.

It stands to reason that our ability to *measure* gain or loss must be considerably better, perhaps 0.01 dB or so. Further, in a large ATE system, measurement error has many components, so it is in our interest to identify and understand error sources that contribute as little as 0.001 dB.

If less accuracy were required, full-channel gain (or loss) would be easy to measure: just apply an audio signal and measure the audio output. But the need for precision complicates the picture, and with codec channels there are several factors to consider.

First, "gain" is not truly meaningful unless the input and output waveshapes are the same. Strictly speaking, this is impossible in codec channels because of quantization distortion and other errors.

Gain is accordingly defined as the ratio of the amplitude of the fundamental component of the channel output to the amplitude of the (pure) input signal, and is expressed in dB. The ideal channel has unity gain ratio (0 dB), so dB gain is synonymous with gain *error*, also in dB.

Gain tracking measures the gain error (in dB) as a function of the absolute input level (in dBm0). It is an excellent indicator of the complementary symmetry of the encoder and decoder "laws."

Figure 12.7 shows the analog concept for measuring channel gain or loss. The bandpass filter isolates the signal fundamental. In a DSP system, the filter and AVM functions, are provided by a software Fourier voltmeter (FVM), typically by the DFT, rather than by the FFT.

To measure half-channel *decoder* gain in a DSP test system, the decoder is driven by a digital bit stream, which is synthesized to simulate the output of an ideal encoder. The analog output is digitized and mathematically analyzed.

To measure half-channel *encoder* gain, the encoder output bit stream is typically converted on the fly to parallel words, and collected in the *receive* memory until a vector of N words has been produced. This vector is then linearized (decoded) either by look-up hardware (RAM or ROM), or by a software algorithm (e.g., MUDEC). The result is a vector of N 16-bit integers or 32-bit floating-point (FP) numbers. The decoded words have 13 bits of resolution.

Choosing the Test Frequency

A most important consideration in sinusoidal codec testing is the tone frequency, F_t. For analog telephone systems the frequencies of 800 and 1000 Hz have long been standard for gain tests. Neither is permitted in codec testing, however, because each happens to be a submultiple of the codec sampling, or *frame*, rate, 8000 Hz.

The problem is best understood in terms of the unit test period (UTP), the shortest period common both to F_t and to 8000 Hz. If F_t is a submultiple of 8000, the UTP contains just one cycle. Prolonging the test beyond this *will not add new information; the same few code levels will be exercised again and again.* At 1000 Hz and the worst-case phase,

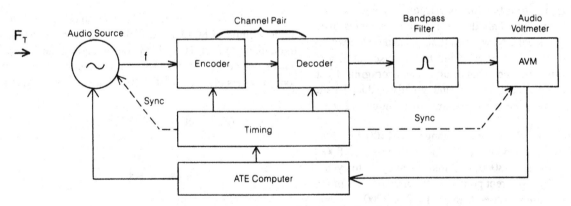

Figure 12.7: Test fixture for full channel gain (analog concept)

only two levels will be tested in each polarity. At 2000 Hz, only one level need occur!

Figure 12.8 illustrates the worst-case phase for these two frequencies. Regardless of the phase, however, the sampling is complete in 1 cycle.

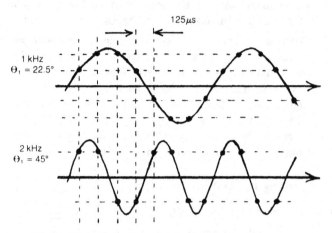

Figure 12.8: Sampling a submultiple test frequency

While these test conditions are obviously not suited for evaluting the transfer curve, they might seem to offer a quick way of testing specific codes. As we saw in Chapter 9, however, ADCs cannot be meaningfully tested by applying single voltages. In mu-law or A-law encoding, in fact, each quantization step is so large—at least 3 to 6 percent of the sampled voltage—that the *uncertainty* of quantization is greater than the permissible gain error! In other words, a perfect codec channel may appear to fail the test limits if tested at 1000 or 2000 Hz!

The solution is to offset the tone frequency and *extend, not repeat,* the UTP. This distributes many samples over many different levels. Now, even though each code may be exercised a number of times, it will be exercised at different levels within the step. This results in an averaging effect for quantization induced error, and greatly reduced gain uncertainty. The principle is illustrated in Figure 12.9.

In this example, F_t has been changed to 1500 Hz. The common, or *unit*, period is 3 cycles or 16 samples. At 8000 s/sec, this becomes 16/8000, or 2 ms. From cycle to cycle, the samples fall at different points on the sinusoid, and more encoding levels are exercised than if a 1000 or 2000 Hz input were repeated three times.

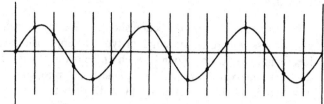

Unit Test Period

Figure 12.9: Sampling a non-submultiple test frequency

In practice, UTP is extended to a much larger number of cycles—to 51, 101, or more—to provide almost continous distribution at the outer (and most significant) code levels.

AT&T gain measurements are made with tone frequencies in the range of 1004 to 1020 Hz. With $F_s = 8000$ s/s, these produce vectors whose UTP contains 2000 and 400 samples, respectively. Since these are not powers of 2, the FVM function is provided by the DFT (FVM) operation, not the FFT. 1004 Hz is the preferred frequency for characterization and standards calibration. 1020 Hz is acceptable for other tests, if the system is coherent.

Example 12.1: Let $F_t = 1020$ Hz. Then

$$\frac{1000}{8000} \rightarrow \frac{102}{800} \rightarrow \frac{51}{400} = \frac{M}{N}$$

The UTP thus contains 51 cycles of the test tone and 400 samples. Its duration is $N/F_s = 50$ msec.

European versions of AT&T tests often employ test tones in the vicinity of 800 Hz. At 820 or 1020 Hz, the primitive frequency (Δ) is 20 Hz, which does not admit the European power frequency, 50 Hz. As a consequence, international codec test sets often use 810 or 1010 Hz as the test frequency.

Example 12.2: Let $F_t = 810$ Hz; then

$$\frac{810}{8000} \rightarrow \frac{81}{800} = \frac{M}{N}$$

The UTP is thus 100 ms. The reciprocal Δ equals 10 Hz.

Frequency Distribution

Quantization distortion adds unwanted energy to the spectrum of the reconstructed channel output. If this noise energy is added to any of the spectral components important to a test—such as the component at F_t when making a gain measurement—then the measurement will be in error.

The *range* of this error is easy to determine in advance of a test. But the distribution and phasing of the noise spectrum are extremely sensitive to encoder gain and to the phase of sampling, and there is enough variation in these two parameters from device to device that the *amount* of error in a particular measurement is uncertain. It is accordingly to our advantage to reduce the range of error, which we can do by choosing test conditions that shift the noise frequency components away from the desired components.

Quantization noise is periodic over the UTP, as are the test signal and the sampling frequency. The digitized sample set is thus essentially periodic, and the reconstructed test signal has a discontinuous, or line, spectrum. The frequencies of interest occupy only a few lines, whereas noise can occupy any and all lines that are numerically valid. The trick is to offer many locations for noise components to appear, thus diluting the spurious energy that falls into any one line.

This is impossible when F_t is a submultiple of 8000, because $M = 1$: *The noise is periodic with F_t and the noise occupies the same lines as F_t and its harmonics.* At $F_t = 1000$ Hz, only three lines are significant: F_t, $2 F_t$, and $3 F_t$.

Making $M > 1$ opens up a new set of possibilities. We can create as many line locations as we desire within the passband, evenly spaced by the frequency increment. $\Delta = F_t/M$. Figure 12.10 compares the spectrum for $M = 1$ (at 1000 Hz) with that for $M = 51$ (1020 Hz).

There are several points of interest:

1. In the 0 to 4000 Hz band, there are N/2 possible lines (disregarding bin zero), spaced apart by Δ.

2. Δ is the reciprocal of the UTP. In Figure 12.10a, $\Delta = 1000$ Hz; in Figure 12.10b, $\Delta = 20$ Hz.

3. At $F_t = 1000$ Hz, four lines are possible; but if N is even (it is eight in this case), the quantization is symmetrical and even harmonics are excluded. *All* the intrinsic quantization energy must therefore fall into the lines of F_t and $3 F_t$.

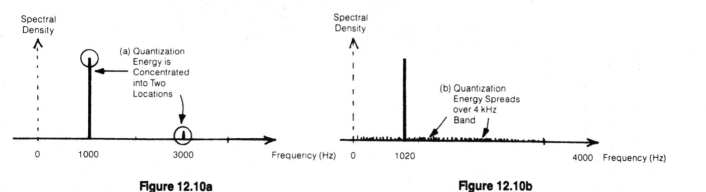

Figure 12.10a

Figure 12.10b

Figure 12.10: Frequency spectrum of the reconstructed channel output at a) 1000 Hz, versus b) 1020 Hz. Perfect mu-law encoding and decoding is assumed.

4. At $F_t = 1020$ Hz, there are 200 bins plus DC. A perfect CODEC will fill alternate lines; a real one may fill all lines (to some extent). Either way, the noise has far more places to appear than when $F_t = 1000$ Hz and the error at F_t or $3 F_t$ is greatly reduced.

5. There are two "fundamentals," the fundamental component of the *Fourier series*, (Δ), and that of the *test frequency* ($M\Delta$). For distinction, Δ is called the *primitive* frequency.

6. Gain measurement by classical analog methods calls for a bandpass filter turned to F_t. Bandwidth and settling time are inversely related, however, and a reasonably fast filter will also let through one or more components on *either side of F_t*. F_t thus acts as a carrier frequency accompanied by side-bands, and the composite signal is *amplitude modulated*. Coherent DSP systems avoid this problem by using Fourier filtering to isolate the component at F_t, while avoiding classical filter settling time.

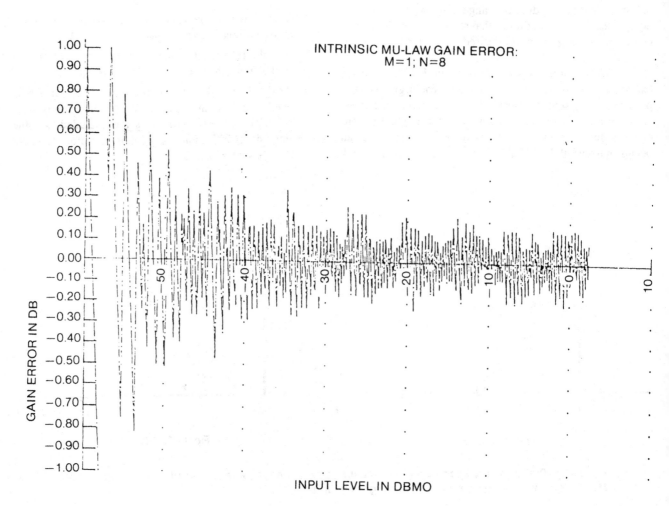

Figure 12.11: Mu-law error

Intrinsic versus Extrinsic Error

At 1000 or 2000 Hz, the gain measurement error intrinsic to mu-law or A-law encoding is large enough to mask some of the manufacturing or *extrinsic,* errors. Figure 12.11 plots the intrinsic error at 1000 Hz, where N = 8.

If M and N are increased, the error becomes smaller. With inputs above −30 dBm0, the peak error in dB is roughly ±/3N when N < 400. For larger N, the error is essentially constant.

The foregoing applies to coherent measurements only. For any N, noncoherent measurements are subject to a wider range of error (uncertainty), and the gain error does not "bottom out" until roughly N = 2000. It is no coincidence that the AT&T tone frequencies are those that produce 400 to 2000 samples. N = 400 is sufficient with coherence only, while N = 2000 is sufficient in all cases.

Why should the error not continue to drop as N is made larger? It is true that the number of possible spectral lines grows greater, but it turns out that the intrinsic mu-law or A-law quantization energy falls only in certain slots. It does not continue to spread.

Figures 12.12 and 12.13 illustrate this limiting effect. Each plot shows the theoretical gain of a perfect mu-law channel as the input level is varied from −60 dBm0 up to +3 dBm0, in 0.1 dB increments. Figure 12.12 is plotted for an *infinitely long* UTP at each level. Figure 12.13 is plotted for F_t = 1020 Hz, where N = 400 samples per input level. The results are nearly identical.

Is there any advantage in using N > 400 with coherence? Yes, if very detailed characterization is to be done. One benefit is that the curve becomes virtually independent of signal phase relative to encoder clock time. At N = 400 or below, the *fine structure* of Figure 12.13 will change slightly with phase. At N = 2000, however, there is no appreciable change.

Transmission Parameters

Loss Variability

A CODEC pair is permitted to exhibit any specified gain, so long as every pairing of that type is consistent. The telephone switch designer can thus provide compensating attenuation elsewhere in the switch circuitry, which uniformly adjusts the overall gain of the switch to 0 dB.

Loss variability is a measure of the consistency of gain, *unit to unit.* It is the gain deviation of a specific connection, compared with the average for that type. The AT&T requirement for 99 percent of all pairing is ±0.5 dB.

Gain Tracking

Whereas loss variability measures the channel-to-channel variation, *gain tracking* measures the deviation in gain of an *individual* channel as the input level is varied. The reference gain is the value obtained at 0 dBm0 (AT&T) or −10 dBm0 (CCITT).

In production, only a half-dozen or so levels are tested. Far more are used in characterization. Figure 12.14 is a characterization plot of a commercial pair tested at 1020 Hz (N = 400). The dotted outline shows the AT&T channel *objective,* and the solid outline shows the *requirement.*

Attenuation Distortion

Another aspect of gain is its variation as a function of frequency, known in telephone terminology as *attenuation distortion.* Since encoding and decoding take place at a constant 8000 Hz rate, the parameter is not so much a measure of the codec ADC or DAC as it is of the filters.

Attenuation distortion would be a routine test were it not for the fact that many codec filters are *sampling* filters and produce image frequencies well into the 200 kHz region. If the filter is integral to the codec, the problem is further compounded by the additional sampling and quantization of the encoding process. High-speed testing of integral filters involves the rules of imaging and undersampling that were given in Chapters 3 and 4.

Signal-to-Distortion

Because of amplitude quantization, even a perfect mu- or A-law channel introduces waveshape distortion into the reconstructed signal. A real channel introduces even more. To evaluate the magnitude of distortion, the total noise accompanying a test signal is measured, relative to the amplitude of the test signal. The ratio is given in dB, and is termed *signal-to-total distortion* (S/TD). The analog form of the fixture is shown in Figure 12.15.

S/TD measurement is preceded by the application of either a "C-message" weighting filter for mu-law, or a "psophometric" filter for a A-law. These are similar; each approximates the subjective effect of noise on the human ear as heard through an American (or European) handset. C-message weighting is shown in Figure 12.6.

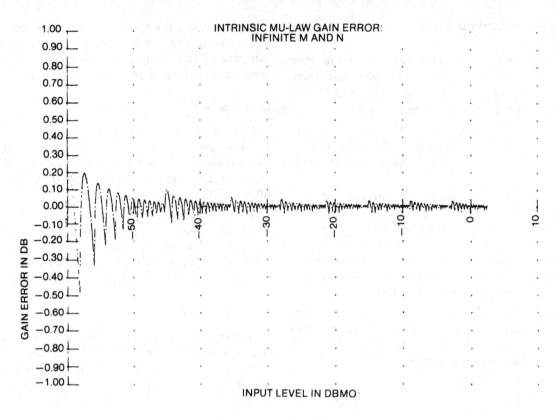

Figure 12.12: Mu-law error

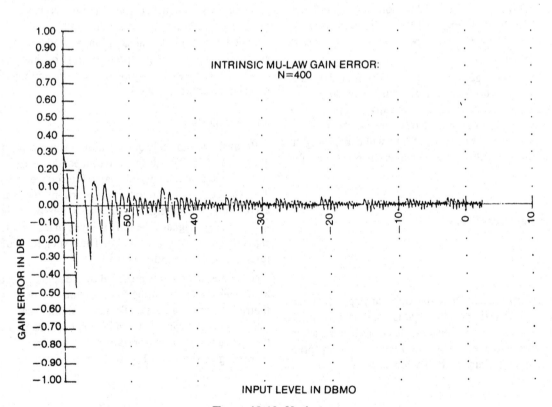

Figure 12.13: Mu-law error

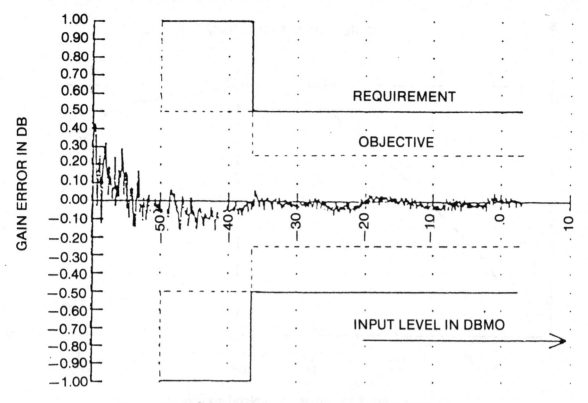

Figure 12.14: Gain tracking plot of a commercial CODEC pair at 1020 Hz

To determine S/TD by the conventional analog approach, the fundamental components of F_t are measured by using the same bandpass filter as for gain. Then a second measurement is made by using a complementary *notch* filter. In a DSP-based system, the equivalent operations are performed mathematically.

In production, this two-part test is repeated at several different input levels. For characterization, we would use many levels and obtain a nearly continuous plot as shown in Figure 12.17. The curve shown is for a perfect mu-law channel with weighting. The channel requirement is shown in the outline.

If F_t is chosen so that N is a power of 2, S/TD can be measured by the FFT operation. The FFT is applied to the digitized vector, and the spectrum is multiplied by another vector that follows the C-weighting or psophometric curve. The signal power is measured, and, then, by the sum of squares (SSQ) operation, the remaining power is computed.

If N cannot be a power of 2 (e.g., at $F_t = 1020$ Hz), the DFT or FVM operation can be used to measure the signal power, and the cosine-sine information used to *reconstruct*

a pure sinusoid of correct amplitude and phase. This sinusoidal vector is subtracted from the original time vector. The result is the *residual*, or *distortion vector*, corresponding to the output of the notch filter. The RMS of the residual is the total distortion voltage.

When N does not permit use of the FFT, spectral filtering can be applied to the residual by a finite-impulse-response (FIR) filter. Alternately, a simple correction factor can be applied that raises the S/TD by the amount of an ideal filter

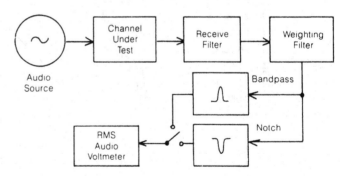

Figure 12.15: Signal-to-total-distortion measurement (analog concept)

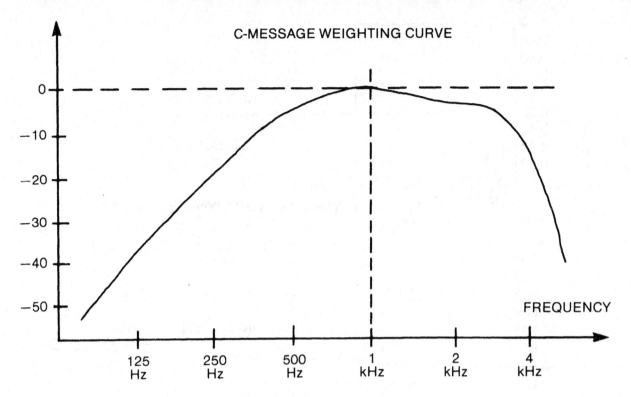

Figure 12.16: C-message weighting curve

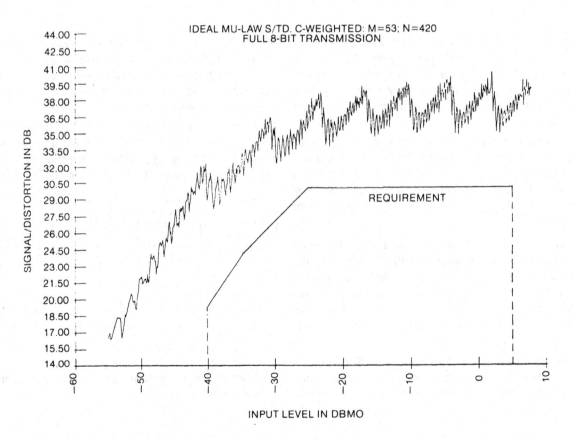

Figure 12.17: Signal-to-total distortion of ideal mu-law channel at 1020 Hz, C-weighted

to Gaussian noise: 2.9 dB for C-weighting, and 3.6 dB for psophometric (CCITT).

Idle Channel Noise

If there is no input signal to a CODEC channel, the channel will be placed in the *idle* state. The encoder continues to sample at 8000 s/s, of course, and its internal noise and auto-zero loop behavior cause it to transmit a somewhat random idle pattern. The decoder converts this to analog noise, adding its own modifications. The final signal is the *idle channel noise*.

The difficulty of measuring ADC idle noise was discussed in Chapter 9, where it was noted that the input should not actually be "grounded" but activated by a keep-alive or dither signal. Each of the various telephone administrations has a somewhat different approach to idle noise. In one approach, a small sinusoid is applied, which stays in the linear region (chord 0), and the noise is measured. In another approach, random noise is used. It is best to consult the standards that apply to the specific DUT.

Nonlinear Distortion

The signal-to-total distortion test does not distinguish between distortion introduced by intrinsic nonlinearity (mu-law A/D quantization) and distortion produced by manufacturing imperfections in the transfer curve. To get a more complete picture, the codec is subjected to further testing that responds in large part to just the classical aspect of nonlinearity. Generally, this is one or both of two closely related tests: intermodulation (IM) distortion and harmonic distortion.

In either test, the key to separating the two kinds of distortion is in the distortion spectrum. Quantization noise is fairly widespread, and tends to occupy most of the spectral slots. Nonlinear distortion shows up in just a few positions. Unlike the notch filter method that transmits all distortion, the filters here must be selective, bandpass units. With DSP, only certain Fourier bins will be examined.

Harmonic Distortion

Low-order nonlinearity in a transmission path manifests itself spectrally as increased energy at $2 F_t$ and $3 F_t$. Lack of symmetry (e.g., where the plus and minus gains are unequal) shows up as second harmonic distortion. Symmetrical nonlinearity causes third harmonic distortion. Higher harmonics are rendered nil by the rolloff of the receive filter when $F_t > 1000$ Hz.

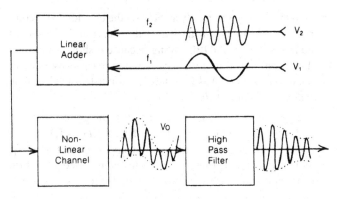

Figure 12.18: IM distortion waveforms

Harmonic distortion (HD) is not a separate CCITT or AT&T test, but a subset of the more general *single frequency distortion* tests (AT&T) or *spurious frequency tests* (CCITT). These limit the allowable amplitudes of single lines regardless of exact location.

Intermodulation Distortion

IM distortion (IMD) testing employs two or more sinusoids and produces a richer variety of distortion products. In theory, each of these products can be deduced from a precise harmonic analysis, but in a band-limited circuit, such analysis is impractical.*

Multitone techniques were discussed in previous chapters, but it is helpful to review the IM concept. Suppose we linearly add two test frequencies, f_1 and f_2, and pass the sum through the channel under test (Figure 12.18).

If the channel is nonlinear, its small signal gain will vary with input voltage. As the low frequency (V_1) "carries" the high frequency (V_2) into different gain regions, V_2 will be *amplitude* modulated by V_1. The amount of modulation can be determined by separating the modulated high-frequency component (V_0) from V_1 with high-pass filtering and then by examining the envelope of the resultant. The depth and shape of modulation indicates the amount and kind of nonlinear distortion.

This simple visualization is inadequate for thorough analysis but many commercial audio IM analyzers do no more

---

*An extreme example is a circuit whose bandwidth is less than one octave. There is no appreciable harmonic distortion, regardless of the degree of nonlinearity. *Multitone* modulation products can easily appear within the band, however.

than this. They just rectify and detect the envelope and compare its RMS value with that of the "carrier." This is of no real value in CODEC testing because it shows only *total* distortion, something we have already learned by simpler means. For the CODEC channel, we must instead look at *individual spectral lines* in V_0.

Second order IMD is characterized by the presence of *sum* and *difference* frequencies, $(f_2 + f_1)$ and $(f_2 - f_1)$. *Third-order* distortion is characterized by components at $(2f_2 + f_1)$ and $(2f_1 \pm f_2)$. If three or more tones are used, the third order also makes a "triple beat" of the form, $(f_1 \pm f_2 \pm f_3)$.

The CCITT IM test is a two-tone test that looks primarily at third-order distortion. The frequency regions are specified, but not the exact frequencies.

AT&T specifies a *four-tone* test that distinguishes both second- and third-orders. The tone frequencies are specified, as are the bins to be analyzed. (See Reference 3 at the end of this chapter.) Both kinds of IM tests can be done in DSP-based systems using either the DFT (FVM) or FFT for analysis, according to the nature of N.

Other Transmission Parameters

There are many other performance tests required for CODEC testing; the ones listed above are merely representative. For a more complete list, the reader should consult the references listed at the end of this chapter.

Half Channel Encoder Testing

In the full channel case, all transmission tests have a common theme: one applies an audio signal to the input, and analyzes the audio (VF) output. But the really meaningful testing must be applied to the channel *halves*, where one of the two signal ports is digital. DSP methods are by far the most practical for half-channel testing.

It is convenient to think of a half-channel test fixture as a full channel; that is, as a channel in which the half not under test has been replaced by a reference unit: an essentially error free substitute.

In half channel encoder testing, the appropriate audio signal is applied to the input. The encoder acts as its own digitizer, providing a serial (PCM) bit stream output. The fixture is like the ADC fixtures previously seen, except that there is a hardware serial-to-parallel converter so that the vector formed has conventional (parallel word) form.

A mu-law or A-law encoded vector has 8-bit words. Before this can be analyzed, it must be decoded (linearized) into words with 13 bits of linear resolution. Decoding can be done in hardware (ROM look up) or in software (MUDEC or ADEC operations). In software decoding, the syntax might look like this:

$$\text{MAT X} = \text{MUDEC (M)} \qquad (12.9)$$

where M is the mu-law (integer) vector and X is the decoded vector. Each element of X is expressed in normalized mu-law units, with a range of -8159 to $+8159$.

Decoder Testing

In the case of decoder testing, the substitute encoder is easy to construct. For any given audio input, the digital response

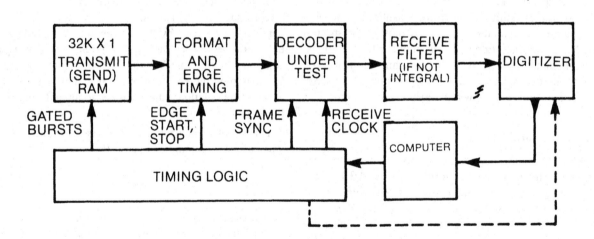

Figure 12.19: Computer-based decoder test fixture

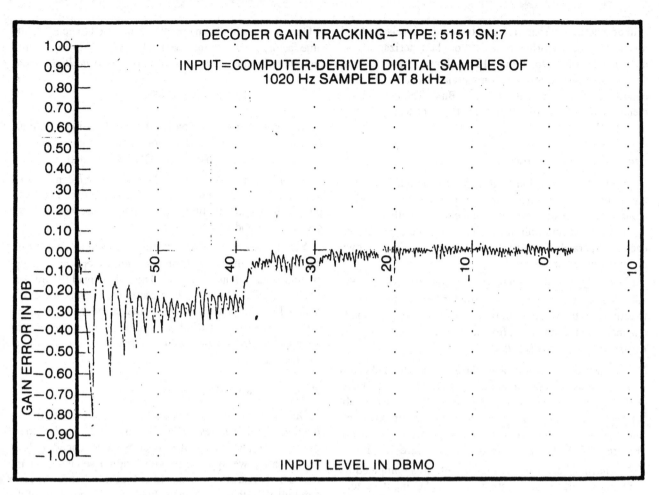

Figure 12.20: Decoder gain tracking

of an ideal encoder is completely determined and can be computed in advance. We may thus replace the audio source and encoder with an all-digital equivalent, consisting of digital pattern memory and timing logic. Simple systems may store a repertoire of precomputed patterns in ROM or PROM, while large systems use a *transmit* RAM (Figure 12.19).

The sequence that simulates the audio source and the encoder sample-hold function is generated by a few high-level commands in DSP BASIC. Let line 10 establish the dimensions of any vectors we may need. For a sinusoid at 1020 Hz, we then write

    20   FOR 1 = 1 to 400
    30   A(l) = E * SIN (45.9*(1 − 1) + T)   (12.10)
    40   NEXT 1

E represents the amplitude of the input sinusoid and is expressed in normalized mu-law or A-law units. The fre-

quency is given implicitly by the argument of the sine function. In this example, F_t = 1020 Hz, and the angular increment is 360 * 1020/8000, or 45.9 degrees. T represents the phase, in degrees. At the end of the FOR/NEXT loop, 400 floating-point samples, A(l) through A(400), will have been generated and stored collectively as vector (matrix or "MAT") A.

Next we encode according to mu-law or A-law:

$$50 \text{ MAT M} = \text{MUCODE(A)} \qquad (12.11)$$

Line 50 applies the mu-law decision levels to each numerical element in A and places the resultant 8-bit code words in the corresponding positions of vector M. M is thus an integer vector whose dimension is 400. Line 60 loads all of the code words of M into the outboard RAM, while other lines start the pattern and initiate the measurement.

In production, perhaps a dozen or so input levels are used for most of the performance tests. For characterization, far more levels are tested. Figure 12.20 shows a gain tracking characterization curve of a commercial decoder, taken at input levels from −60 dBm0 to +3 dBm0. The input increments are 0.1 dB, giving 630 levels in all. The test frequency is 1020 Hz.

Removing Intrinsic Error

Quantization distortion (and thus intrinsic mu-law or A-law error) is a mathematical property of the encoder ADC alone, not the DAC. In a full-channel test, in other words, quantization occurs only once. If we test the encoder and decoder separately, however, each half seems to have its own quantization error.

For production testing this is not a problem, because the AT&T and CCITT templates allow for this. But if you try to estimate full-channel performance by combining two half-channel transfer functions, for example, the result will have *excess quantization distortion.*

Quantization does not take place within the DAC, of course. It exists in the digital input pattern. To see the DAC performance independent of pattern quantization, we can subtract the intrinsic mu-law gain tracking error already computed (Figure 12.13) from the actual gain tracking curve (Figure 12.20). The result is shown in Figure 12.21.

The intrinsic error of the digital pattern is computed by applying the decoding function to the DAC input vector,

$$\text{MAT } B = \text{MUDEC } (M) \qquad (12.12)$$

and then applying the DFT (or FVM) to measure the amplitude at F_t. This operation is repeated for each different input level—typically a half-dozen levels for production testing, or several hundred for characterization.

Other Decoder Performance Tests

The principle illustrated for gain tracking applies to all other decoder transmission tests: Apply a digital sequence that simulates the output of a perfect encoder, then measure the decoder response through the appropriate measurement filters. One can easily generate any waveshape by adding different sinusoids or by using a random or pseudo-random function for noise-based testing. The general expression for each sinusoidal term required is

$$\ldots + E * \text{SIN } ((l - 1) * 360 * F/8000 + T) \qquad (12.13)$$

Once a series has been completed and used to generate the time factor, the vector may be quickly rescaled to any desired amplitude by the matrix operation of *scalar multiplication:*

$$\text{MAT } B = <\text{expression}> * A \qquad (12.14)$$

The multiplier expression is usually a function of dBm0. To run a detailed characterization curve, it may be successively changed as a part of a FOR NEXT loop.

When N is a power of two, the *inverse FFT* can be used to create complex time series from spectral information. Sample programs for multitone and pseudo-noise waves were given at the end of Chapter 6.

You can see that decoder testing is essentially no different in concept than full-channel testing, except that we supply the digital stimulus directly. The full-channel fixture concepts outlined for other transmission parameters are easily adapted to half-channel encoder and decoder. There are a few tests peculiar to half-channel parameters, however, and we conclude this chapter with two that apply specifically to the decoder

Digital Milliwatt Testing

One decoder test that has no full-channel counterpart is the digital milliwatt test. This establishes the *absolute* magnitude of output voltage as a function of the encoding law. Specifically, we apply a periodic digital sequence to the decoder representing a sinusoid at 0 dBm0 (hence the term "milliwatt"), then measure the fundamental amplitude of the *reconstructed* output sinusoid. To generate such a digital sequence shown earlier, the value of E in line 30 (Equation 10.10) would be set to 5663 mu-law units, or 2853 CCITT A-law units. (These relationships were given earlier in this chapter.)

This is the *only* test permitted by the CCITT to be run at 1000 Hz. 0 dBm0 is, in fact, defined for both encoding laws so that if T = 22.5 degrees, the quantization error in the fundamental component will be rendered nil at 0 dBm0 when N = 8.

Idle Noise

In a full channel, most of the idle noise comes from the encoder. The trick is to isolate the smaller contribution of the decoder. For A-law, the CCITT recommends the application of an alternating input pattern—the two center codes—to simulate the behavior of the idealized encoder auto-zero circuit. The "canonical" pattern has energy only at 4000 Hz and above, which is greatly attenuated by the receive and

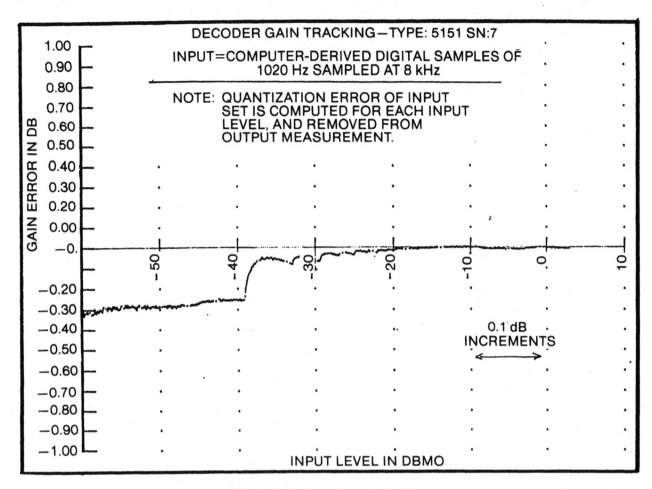

Figure 12.21: Decoder gain tracking

weighting filters. Presumably, any excess energy, particularly in-band, is then a direct measure of *decoder* imperfections.

Mu-law is treated differently. Alternation of the two center codes produces no output from a good mu-law DAC, and in real life is an inadequate test of zero-crossing errors. AT&T recommends a decoder drive pattern that simulates an encoder driven by a noise generator whose amplitude is adjusted to produce +19 dBrnC0 in a reference channel. (This is not the same as the random digital pattern of +19dBrnC0.) A real decoder driven by this pattern should not exceed +20 dBrnC0 at its output.

References for CODEC (PCM) Telephone Standards

1. "Transmission Parameters Affecting Voiceband Data Transmission-Measuring Techniques," *Bell System Publication 41009*, AT&T, Basking Ridge, N.J., May 1975.

2. *CCITT Yellow Book*, Volume III-3, International Telecommunication Union, Geneva, 1981. PCM parameter information is repeated in the subsequent edition, the *Red Book*, 1984.

3. "Local Switching System General Requirements (LSSGR)," *Bell System Publication 48501*, Volume 1, Section 7.4, AT&T, Basking Ridge, N.J., Dec. 1980

Chapter 13
Selected Reprints

AUTOMATED MEASUREMENT OF 12 TO 16-BIT CONVERTERS

MATTHEW MAHONEY
Staff Consultant

LTX CORPORATION
145 University Ave.
Westwood, MA 02090

ABSTRACT:
Direct measurement of Analog-to-Digital
(A/D) and Digital-to-Analog (D/A) conversion parameters is an exacting task
at levels of 12 bits and above. This
paper describes five elements of an indirect computer-based technique which
can produce precise voltage measurements from fixtures and instruments
of ordinary accuracy.

INTRODUCTION

How is it possible to test precision
A/D and D/A converters in production?
The question is not frivolous, for
modern converters are often capable
of higher accuracy than the instrumentation found in automatic test systems.

It is not enough to propose more accurate
instruments. A 16-bit converter, for
example, has a nominal step size of 15
parts per million. To keep the measurement error an order of magnitude below
this demands cumulative short and long-
term system errors below 2ppm! It is
unrealistic to expect that this kind of
laboratory precision can be maintained
at a test socket two meters removed from
the system voltmeter, and at speeds high
enough for production.

The problem is akin to that of the
diamond-cutter, whose tools do not have
edges as hard as the diamond he seeks to
cut. The answer of course lies in
cleverness and technique, not in direct
assault.

So it is with testing converters of 12
bits and more. By exploiting the structure of these devices, the right technique
allows ordinary instruments to make extra-
ordinary measurements. To succeed,
approaches of this kind draw on two
properties: indirect measurement, in
which no answer is obtained by a single,
absolute measurement, and "bootstrapping",
in which each stage of the procedure
builds upon the accuracy of the preceding
one.

In this article we show some of the
elements of indirect D/A and A/D measurement, particularly those that utilitze
the computational ability of a modern ATE
system. To better appreciate the utility
of such procedures, let us first review
the shortcomings of direct measurement.

Direct Mapping

Consider the testing of a digital-to-
analog converter (DAC). Ideally, a DAC
serves as a one-for-one mapping device
between the digital and analog domains.
There is a finite, well-delineated set
of input states, and for each state there
is an unique, corresponding value of output voltage or current. Although we are
frequently shown DAC graphs with steps or
staircases, the graphical map is really a
point set (Figure 1).

In principle, the DAC exhibits no uncertainty in either forward or backward
direction. The horizontal spacing is by
definition uniform and repeatable for all
like units under test. Device errors
manifest themselves as vertical displacements of any or all points, either stat-
ically or dynamically. For those cases
where dynamic errors are unimportant, an
input-output map tells us what we need to
know.

The test method seems obvious. Why not
simply measure the vertical position of
each of the transfer points, and compare
it to the design limits? Why not, in
other words, made a map of each unit?

Reprinted from 1981 IEEE Test Conference, 1981, pages 319-327. Copyright
© 1981 by The Institute of Electrical and Electronics Engineers, Inc.

Some 8-bit devices are actually tested this way. But direct mapping is inappropriate for precision converters on several counts.

Time

First, there is the sheer weight of numbers. A 12-bit converter has 4096 states; a 16-bit unit has 65,536. The time required to measure and algebraically manipulate the full set of values is prohibitive in production, running from many seconds to many minutes.

Fortunately, much of the information in a full point set is redundant, and we need not look at every element except for characterization. The trick is to choose the smallest subset that does not omit any vital information, yet at the same time facilitates the bootstrapping process.

Accuracy

Second, to measure the map elements directly — that is, to use single, absolute measurements — demands an unnecessarily high accuracy. This is amply illustrated in measuring those converter parameters that are derived from the non-uniformity of output increments, or "steps". Such parameters are usually specified to $\pm 1/2$ or $\pm 1/4$ LSB, and it is ordinarily sufficient to restrict the measurement error to $\pm 1/10$ LSB.

In one sense, this is easy: just measure each interval to 10% accuracy. But we cannot measure the step itself directly, because the two voltages that define it occur at different times. The trouble starts when we try to calculate step size from the difference of two absolute measurements.

Let V_x and V_y represent the smaller and larger of two adjacent voltages, respectively. What we want is the difference, $V_y - V_x$. What we actually obtain is

$$(V_y + C + e_y) - (V_x + C + e_x)$$

Here, C is the repeatable component of measurement error, while e_x and e_y are the non-repeatable components. The latter include the tracking error of the voltmeter between the two voltages, short term drift, and random, noise-induced errors.

For simplicity, assume that e_x and e_y are bipolar and have the same range of variability. If we denote the peak error simply as e, then the worst-case calculation is

$$V_y - V_x \pm 2e$$

At 14 bits, the nominal step size is 61 ppm, roughly 600 microvolts in a 0 to 10 volt converter. To limit measurement error to one-tenth of this, we set 2e equal to 60µV, giving an error limit, e, of 30 µV.

In direct measurement, the voltmeter range must at least equal V_y. But on the 10 volt range, 30µV is only 3 parts per million. This is a very hard way indeed to make a 10% measurement!

Indirect techniques avoid this trap by avoiding measurements where the absolute values are large compared to the differences to be taken. By bootstrapping several stages of medium-accuracy indirect measurements, we can easily bring the measurement errors below the random noise of the DUT.

Noise and Measurement Time

Random noise has a far greater effect on measurement accuracy than is generally appreciated. In the above example, if all other errors are rendered nil, the goal of a 30µV error limit would be reached and occasionaly surpassed by 10µV RMS of Gaussian noise. This is the amount produced in a 500kHZ band by thermal agitation in a single, perfect 10K ohm resistor!

Bandwidth provides a simple means of noise control, and the fixture we will show employs controllable bandwidth. Although indirect measurement is primarily aimed at increased accuracy, it also has a secondary benefit of increased speed when the bandwidth is restricted:

1. By reducing non-random errors, it permits more noise for the same error budget, and thus allows greater bandwidth and measurement speed.
2. For a given time constant, low accuracy measurements take less settling time. To read a step to 10%, one must allow only 3 time constants. To obtain the same result by subtracting 3ppm absolute measurements requires 13 time constants for each reading.

Indirect measurement offers other advantages. but we have seen enough to start us in the right direction. Let us now assemble some hardware and software "building blocks" for precision testing. Specific examples are taken from the commercial A/D, D/A test fixture incorporated in the LTX MTS77 system.

I. THE DIFFERENTIAL FIXTURE

Absolute differences suffer largely because the arguments, V_x and V_y are so much larger than their difference. To avoid this problem, we will devise an algorithm that does not require any knowledge of V_y or V_x individually, just the difference. Then we can use an ordinary DC supply to bias the voltmeter and move the measurement baseline, V_b, towards V_x, as shown in Figure 2.

It is not necessary for V_b to be accurate (i.e., to take a known value), but it must be quiet and have good short-term stability. The resolution must be sufficient to place V_b within a few millivolts of Vx; then the two voltages seen by the voltmeter, (V_x-V_b) and (V_y-V_b), will each be in the order of the difference, (V_y-V_x).

While supply accuracy is not mandatory, it saves time by allowing V_b to be placed within range in a single operation. Otherwise, the computer must recursively re-adjust the supply to bring the voltage within range. Figure 3 shows the simplified arrangement for the production fixture. V_b is provided by a local low-noise programmable DC source with 0.1% accuracy and much finer resolution. Because the system voltmeter is remote to the fixture, the difference, $V-V_b$, is first amplified by a low noise instrumentation amplifier (x100) then transmitted over a balanced, shielded line to the voltmeter.

Differential measurement by this technique reduces most of the common voltmeter errors:

1. Amplifier and voltmeter offset cancel in subtraction.

2. Gain error applies only to the difference, which in our case is always on the order of 1 LSB. Even this small error will be made to vanish in the algorithm we use.

3. Voltmeter tracking error (total non-linearity) is a function of the range setting and the location of the two readings. The improvement from range reduction alone is significant, as illustrated by a typical example.

Let V_b fall 4mV below V_y and assume the step to be measured is $600\mu V$. With x100 amplification, a range of 0.5 V is optimum. The LTX voltmeter tracks to better than 0.02% (200ppm) FSR, giving $\pm 100\mu V$ error for the worst case.

Referred to the DUT, this is only $\pm 1\mu V$ or 1/600 LSB.

4. In many voltmeters, the major tracking error is well-behaved (smoothly curved). There, tracking error can be reduced further by keeping the two readings close to each other. We will show a trick for this shortly.

5. The voltmeter and cabling noise after the preamplifier runs around $100\mu V$ RMS. Thanks to the x100 preamplifier, this is effectively only $1\mu V$ RMS at the DUT.

6. The real limit to accuracy in the differential fixture is random noise at the DUT output terminals - the combination of preamplifier, bias supply, DUT, and fixturing noise. There are no shortcuts here, just the application of physical principles.

Noise Considerations

The fixture described has a preamplifier bandwidth of 4 kHz, and (with the DUT bypassed) exhibits and equivalent input noise on the order of $10\mu V$ RMS. But this is deceptive for we are not dealing with Gaussian white noise, but semiconductor noise - popcorn, flicker, 1/f and all. Occasional peaks run over $100\mu V$. The system voltmeter deals with the problem in two ways. To reduce random noise, the voltmeter operates in the Average mode, in which the system computer requests a program-selectable number of successive readings. It discards the maximum and minimum ones, then averages the remainder. Averaging 100 samples in this fashion ensures repeatability of readings at the preamplifier terminals within $\pm 10\mu V$, peak.

Averaging is effective only if the samples are independently affected by noise. For random noise limited to 4 kHz, the maximum useful sampling rate is 8000 samples/second. Averaging 100 samples requires at least 12.5 ms to be fully effective.

Digital crosstalk poses a serious problem because the "noise" is not random, but may add in some pattern not removed by short-term averaging. To deal with this kind of noise, the voltmeter employs the "Slow Average" mode, in which the "slow" modifier causes the digital bus to fall silent for 1 ms prior to a voltmeter sample. In this mode, averaging 100 samples takes a little over 100 ms.

One may tailor the number and timing of samples to fit the precision and throughput needs of the test. With averaging, it is feasible to measure increments with an accuracy equivalent to 1 ppm of the device's full scale. Let us now put these increments to work.

II. THE LINEARITY ALGORITHM

Often the gain or range of a converter is not trimmed until the linearity is first measured and trimmed. We start then with an algorithm for linearity, and moreover, one which must operate without exact knowledge of the electrical span.

Assume initially that the DUT has no superposition error - that is, that each of the n bits contributes a weighted amount of voltage or current to the output level that is independent of the other n-1 bits. In a binary converter, the LSB should contribute 1 unit, the next bit 2 units, the next bits, 4,8,16, and so on.

Let i denote the bit position, from 1 (the MSB) to n (the LSB). And let V_i represent the individual contribution of the ith bit. In the absence of superposition error, any of the other 2^n outputs can be computed by summing the V_i for the active bits and subtracting the offset (the DUT output when no bits are active). Furthermore, one could obtain a rough check on superposition by comparing the sum of all V_i (plus offset) against an actual measurement with all bits active.

But of course this involves large absolute measurements, whereas our fixture is designed only for small increments. In fact, it is really restricted to unit steps of just one form:

$$\uparrow \left.\begin{array}{l} 0001000000 \\ 0000111111 \end{array}\right\} \Delta i$$

Each Δ_i consists of V_i less all lower order V. Given all n values of Δ, the set of V_i is obtainable by algebra, and from this - if desired - a map of all points relative to the device's own scale. For linearity, this is all we need.

The derivation of INL from Δi is based on a few simple relationships:

1. For each V_i, there is a corresponding B_i, defined as the nominal contribution of the ith bit. B_n is thus the nominal LSB.

2. The nominal contribution is a binary fraction of the Full Scale Range (FSR);

$$B_i = FSR/2^i$$

3. The FSR is the contribution of the hypothetical bit beyond the MSB. As this is not measurable, we work instead with the electrical span, FSR - 1 LSB. Assuming that offset is always removed from our calculations, the span is equal to Vmax, the output with all bits active.

$$B_n = Vmax/(2^n - 1)$$
$$B_i = 2^{n-i} B_n$$

4. The difference between the actual bit contribution and the nominal value is the bit error, e_i.

$$e_i = V_i - B_i$$

5. The nominal values are always scaled to the actual Vmax. Thus

$$Vmax = \Sigma V_i = \Sigma B_i$$

6. The sum of all bit errors is therefore zero

$$\Sigma e_i = 0$$

7. Since individual e_i are non-zero, the error set must be bipolar. Furthermore, when the positive errors are grouped separately from the negative errors, the two sums must have the same magnitude.

$$\Sigma(+) + \Sigma(-) = 0$$

8. These two sums define the + and - limits of integral non-linearity. Ideally, these are symmetrical, but measurement error is likely to introduce some inequality. To minimize the effect, the average is usually taken:

$$INL = \frac{\Sigma(+) - \Sigma(-)}{2B_n}$$

9. INL is thus determined from the set of e_i, i.e. the difference of sets V_i and B_i. These in turn will be derived from the set of increments, Δi, provided by the differential test fixture.

10. To ensure that the percentage error in e_i be no greater than that of V_i or B_i, it is essential that all terms be derived from a common set of data. Only the Δ_i for each device will be used; no absolute measurements are permitted.

An All-Differential Algorithm

For any n or i, V_i is always the sum of the immediately preceding unit step, Δ_i, plus binary multiples of all lower-order Δ. The coefficient pattern is the first n numbers of the series.

$$1 + 1 + 2 + 4 + 8 + 16 + \ldots.$$

Figure 4 illustrates its application to a 5-bit converter. By inspection, one can obtain the five equations listed below.

V_1	$=$	$\Delta_1 + \Delta_2 + 2\Delta_3 + 4\Delta_4 + 8\Delta_5$
V_2	$=$	$\Delta_2 + 1\Delta_3 + 2\Delta_4 + 4\Delta_5$
V_3	$=$	$1\Delta_3 + 1\Delta_4 + 2\Delta_5$
V_4	$=$	$1\Delta_4 + 1\Delta_5$
V_5	$=$	$1\Delta_5$

To find the nominal values, B_i, we first find the nominal span, Vmax. This is obtained by summing the equations above:

$$Vmax = \Sigma V_i$$
$$= \Delta_1 + 2\Delta_2 + 4\Delta_3 + \ldots\ldots$$

The B_i are obtained from this by equations (3) and subtracted from V_i. The final equations for e_i follow a pattern best seen in two steps. First, the MSB error, e_1, is given by the fraction

$$e_1 = \frac{(2^{n-1}-1)\Delta_1 - \Delta_2 - 2\Delta_3 - 4\Delta_4 - (2^{n-3})\Delta_n}{2^n-1}$$

The example for n = 5 is given below:

$$e_1 = \frac{15\Delta_1 - \Delta_2 - 2\Delta_3 - 4\Delta_4 - 8\Delta_5}{31}$$

Second, the values of e_2, e_3, etc., are obtained by shifting the numerator coefficients circularly to the right, as shown below.

BIT	MAGNITUDE OF COEFFICIENTS FOR:					DENOMI-NATOR
	Δ_1	Δ_2	Δ_3	Δ_4	Δ_5	
1	+15	-1	-2	-4	-8	
2	-8	+15	-1	-2	-4	
3	-4	-8	+15	-1	-2	31
4	-2	-4	-8	+15	-1	
5	-1	-2	-4	-8	+15	

To complete the algorithm, each of the e_i is calculated; the positive and negative values are grouped separately; then two sums are generated which define the integral non-linearity.

The scheme is competent to measure non-linearity to 3 ppm of full scale, i.e., 1/5 of LSB at 16 bits. Yet no absolute accuracy much better than 1% is required. The preamplifier and bias supply must simply be quiet and have good short term stability.

The Error Budget

There are three elements in the algorithm that contribute to the removal of non-noise errors from e_i. The first is the size of the denominator, 2^n-1. It is twice the size of the positive coefficient and twice the sum of all the negative ones. Thus an error of 1% in just one Δ would cause only a 1/2% error in the computation of D.N.L.

The second element is that for all e_i, the sum of numerator coefficients is zero. Thus if all Δ's are subject to the same constant error, and the error vanishes in e_i.

The third element is seen in the expression for INL, equation 8. If all Δ_i are altered by the same proportional error, the numerator and denominator are altered in equal proportion, leaving the expression for INL unchanged. Thus voltmeter and amplifier gain error have no effect.

The only errors that remain, in fact, are those that independently affect the various Δ_i. Primarily, these have two sources: random noise (which is the ultimate limit) and amplifier/voltmeter tracking error. The latter shows up because the point pairs fall at different locations on the voltmeter transfer curve, and have been deliberately spread apart over the voltmeter range by the preamplifier.

We showed earlier that with a good voltmeter, the effect is trivial. But the second algorithm property allows us an unusual opportunity to reduce even this residual effect. The method is due to Solomon Max.

III. EXTRA-BIT NULLING

To the original fixture we add a current or voltage switch which serves as an additional bit on the DAC under test. (Figure 5) Its contribution, k, is programmed to equal the expected LSB, and is logically connected to the LSB input of the DAC under test.

Recall that our fixture is used only to measure unit steps where the smaller level corresponds to an odd code and the larger to an even code. This causes the extra bit to be added only to the smaller of the pair of voltages. The increment measured by the voltmeter is then the residual

$$d = \Delta_i - k$$

Substituting d into the original equations for e_i splits the numerator into two groups, one containing the variables, Δ_i, and the other containing the constant k. Since the sum of the numerator coefficients is zero, the second group vanishes. Without affecting the calculation, the two voltage readings have been brought close together, and tracking error is greatly reduced. The hardware error is now scaled to d, providing good example of boot-strapping: the more precise the DUT, the more precise the measurement.

IV. THE SUPER-LINEAR REFERENCE

Now we must measure total non-linearity. Most systems at this stage resort to absolute measurements at Vmax, Vmin and intermediate points. But to measure a 16-bit device to even 1/2 LSB takes measurement accuracy of 8 ppm or better. As a consequence, many commercial converters omit the total non-linearity specific ation, or simply quote what the device ought to do.

An all-differential approach would work if our fixture were additionally equipped with a reference or "golden" DAC which was perfectly linear, and whose slope and offset could be trimmed by software to match endpoints of the DUT. (Actually, it suffices to bring them close enough that the separation is within the differential range of the fixture.) The device that serves in this capacity is termed the Super-Linear Reference, or SLR.

We cannot really build a DAC which will stay linear over 10 years at the part-per-million level. But we do have an algorithm and fixture that can measure differential non-linearity to 3 ppm. So this portion can be removed in a computer-based SLR design.

Can we avoid the other part, namely superposition error? Not totally, but a good D/A designer can reduce it to a few ppm in a discrete-component circuit. The significant factor, however, is that unlike other D/A errors, the super-position error is essentially unaffected by component aging or temperature: it is instead a function of the kind of circuit design. The designs that have very low integral non-linearity tend to drift in DNL, but that is quite acceptable to our computer-based SLR.

The SLR need not have as much resolution as the DAC under test. It is only necessary to bring its output close enough to the DUT output to ensure that the difference is on the order of a few LSB's. In fact, it need not truly be linear; but it must be calibrated so well that the position of every one of its 2^n outputs is precisely known relative to Vmax.

In the fixture described herein, the SLR is a 16-bit discrete-component Kelvin (4 wire) DAC with about 3 ppm long-term superposition error. It is frequently and automatically calibrated by the DNL algorithm and the error terms are incorporated into the DUT test program. Because

the major cause of short term linearity drift is internal temperature change, the SLR temperature is monitored by a thermistor, interrogated once per second by the computer. A change of 0.5 degree since the previous calibration iniates the next automatic calibration. A selectable time interval is also provided.

To measure the total non-linearity of the DAC under test, the computer matches the DUT and the SLR endpoints then measures the differences at other program-selectable points. Residual gain or offset differences are removed in software. The voltmeter, as always, sees only the magnified differentials.

The SLR itself has about 7 ppm cumulative track-ing uncertainty: 3 ppm from the calibration, 3 ppm superposition error, and 1 ppm for drift between calibrations. To this we add another 2 or 3 ppm in reading the DUT-to-SLR differential. This brings the practical error in measuring DUT tracking error to 10 ppm FSR.

V. A PRECISION ABSOLUTE VOLTMETER

To this point we have avoided absolute measure-ment. A precision DAC requires at least one such test, if only to measure the accuracy of its internal reference voltage. Our final level of "bootstrapping" turns the SLR to this end.

The principle is over a century old: the potentiometer, or potential-measuring divider (Figure 6). A voltage of known value is applied to a linear divider, which in turn is adjusted until the galvanometer is nulled. The only real distinction is that we do not here require a null, just a small enough difference for our original fixture to be within its optimum range. The SLR serves beautifully as the potentiometer it-self, with a linearity within 7 ppm FSR, while the differential measurement itself adds another 2 or 3 ppm.

The reference, E, is provided by a stabilized voltage source, but it is not the system standard. The division of E into the individual bit contributions is subject to change with temperature and age. For highest absolute accuracy, the automatic calibrations therefore depend not on E, but on a set of reference standards from a rack-mounted, oven controlled, multiple voltage source. The concept is shown in Figure 7.

In keeping with our philosophy, none of the standards is adjustable, for adjustments general-ly mean less stability. Twice a year the standards are measured in a metrology lab, and the absolute errors from nominal are transferred to the system disc, and incorporated into the automatic calibration program.

In simple terms, it is easier to make a 7 ppm linear divider and 3 ppm nullmeter than it is to make a 10 ppm analog-to-digital converter. A linear division is advantageous because absolute full scale _error_ is divided, too, and appears as a percentage of reading. Not so with]0 ppm non-linearity error. It may do its dirty work anywhere in the range. And although no sales literature will direct the reader's attention to the fact,]0 ppm FSR non-linearity error may contribute 30 ppm to a one-third scale reading.

The effect is minimized by keeping the reading in the upper portion of the range. To do this, the SLR output is equipped by a binary-ranged divider, and each range is calibrated independently for absolute magnitude. We have used only unipolar examples, but the SLR is actually a bipolar device. Its actual full-scale ranges are +13V, +6.5V, +3.25V, and +1.625V. The (+) and (-) zones are calibrated independently and the linearity specification applies not to the full span, but to either endpoint. Thus the two most common DAC reference voltages, 6V and 10V, appear at 92% and 83% of full scale, and the 10 ppm FSR linearity appears as 11 ppm and 12 ppm relative to reading. Added to this is the absolute error, which is largely a function of the metrology laboratory and the time since the last updating. Values of 30 to 40 ppm are most likely with 6 months intervals.

SUMMARY

The foregoing techniques illustrate the power of indirect measurement, especially when one measurement is used to "bootstrap" the next. The fixture and algorithm described permit the production measurement of differential non-linearity to 1/3 LSB at 16 Bits; relative accuracy to 10 ppm FSR; and _absolute_ NBS-referenced accuracy to 50 ppm of the reading itself. Yet no element needs much better than 1% accuracy (10,000 ppm). The requirements are low noise, good stability, and occasional access to a laboratory standard.

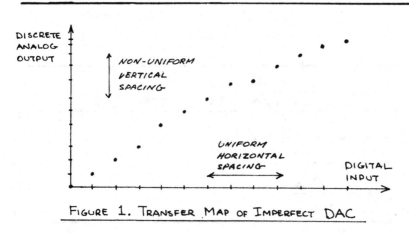

FIGURE 1. TRANSFER MAP OF IMPERFECT DAC

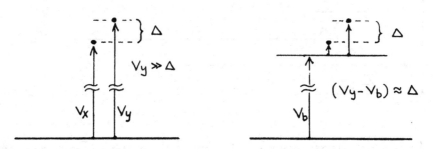

FIGURE 2. ABSOLUTE VS. BIASED DIFFERENCES

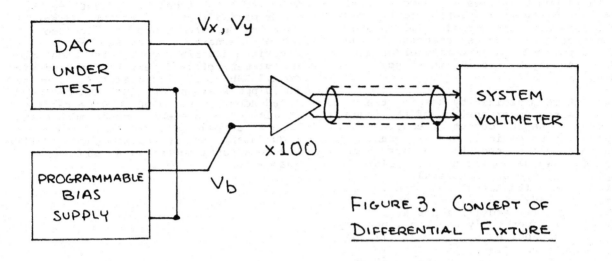

FIGURE 3. CONCEPT OF
DIFFERENTIAL FIXTURE

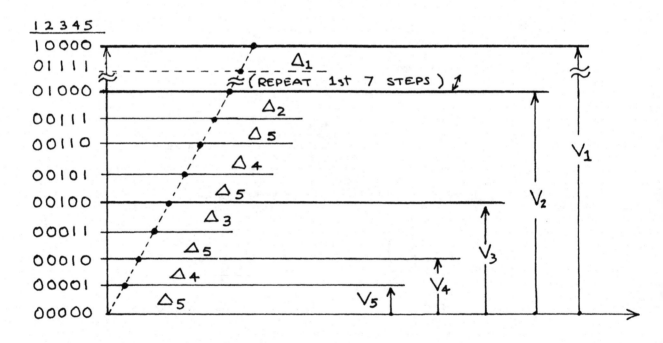

FIGURE 4. SUPERPOSITION OF UNIT STEPS IN 5-BIT DAC

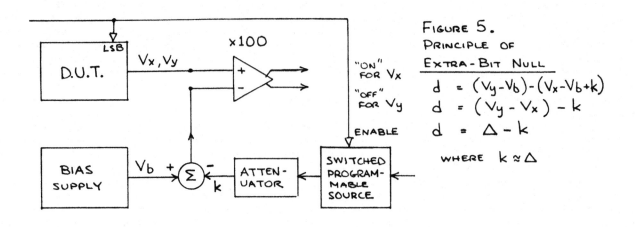

FIGURE 5.
PRINCIPLE OF
EXTRA-BIT NULL

$$d = (V_y - V_b) - (V_x - V_b + k)$$
$$d = (V_y - V_x) - k$$
$$d = \Delta - k$$

WHERE $k \approx \Delta$

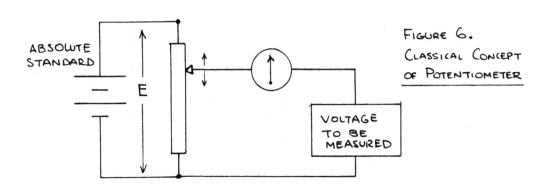

FIGURE 6.
CLASSICAL CONCEPT
OF POTENTIOMETER

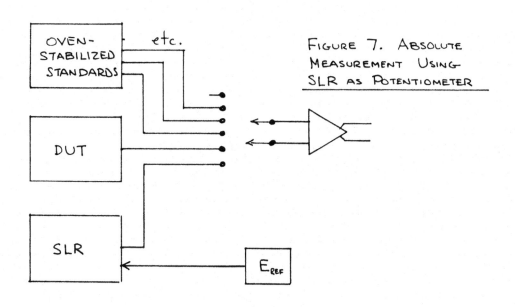

FIGURE 7. ABSOLUTE
MEASUREMENT USING
SLR AS POTENTIOMETER

DSP Measurement of Frequency

Eric Rosenfeld
LTX Corporation
Westwood, Massachusetts

ABSTRACT

Digital Signal Processing (DSP) test techniques combine sampled-data methods with mathematical analysis to make high accuracy analog measurements. This paper presents such a technique for measuring the amplitude and frequency of periodic signals that are not synchronized with the measurement system sampling frequency. This technique is based on conditioning the data with the "Rosenfeld" window prior to discrete Fourier analysis.

INTRODUCTION

In the past, the primary technique available for measuring frequency was to time one or more periods using a hardware counter/timer. This method requires that the signal component being measured be separated from other components by filtering. For example, when measuring the frequencies of telephone DTMF signals, it is necessary to use low and high band filters to separate the tones.

A new technique is now available that can be used for frequency estimation. This technique does not require any special signal preconditioning other than conventional anti-alias filtering. Figure 1 shows a typical DSP measurement configuration.

The new technique takes advantage of the precise mathematical properties of the Fast Fourier Transform (FFT). The FFT can be thought of as a bank of filters. Each filter center frequency is separated from the next by the Fourier frequency (the reciprocal of the time product of the sampling period times the number of samples collected). The shape of the filters is determined by the very nature of the FFT. If a signal which does not fall exactly at the center of one of the analysis frequencies (i.e. a periodic signal not synchronized with the measuring system sample frequency) is filtered using an FFT, the response of each component of the FFT will be dependant upon the shape of the filter. By looking at

several components of the FFT in the region where the largest component is located, it is possible to make an excellent estimate of the frequency of the signal.

The technique in its practical form requires the application of a "window" prior to performing an FFT. If a digitized signal is FFTed without windowing, signals that are not periodic over the data collection interval will produce components at frequencies that are far away from the actual frequency of the signal. In addition, the results of the FFT will be dependant on the phase of the measured signal. If however, the digitized time sequence is multiplied by a suitable "windowing" time sequence, the energy of a signal is localized over a limited range of frequencies, and the distribution in frequency is no longer phase dependant.

CHOOSING A WINDOW

Three different windows were studied, the Hann window, the Blackman window, and a custom window (the Rosenfeld window). The nature of these windows is similar. However they represent different compromises. The Hann window produces a narrower spread of components near the signal frequency, but allows more of the signal to "leak" into frequency components further from the signal frequency. The Blackman window produces a broader spread near the signal frequency, but less leakage at other frequencies. The spread of the Blackman window is about twice that of the Hann window. However, the out-of-band leakage of the Blackman window is 20dB less that the Hann window. The Rosenfeld window is similar to the Blackman window, but it is optimized to minimize side-lobe energy. In practical terms, this means that two signal

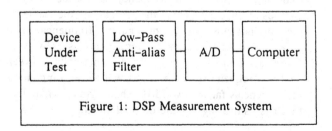

Figure 1: DSP Measurement System

Reprinted from *1986 International Test Conference*. 1986. pages 981-986.
Copyright © 1986 by The Institute of Electrical and Electronics Engineers, Inc.

components must be slightly further apart when using the Blackman or Rosenfeld window. But if the two signals are sufficiently far apart to avoid gross interference, the Rosenfeld window will produce no interaction between components.

THE ROSENFELD WINDOW

When designing a window for a particular application, it is useful to have some understanding how windowing operates in general. When data is collected using sample-data techniques, the continuous analog signal is sampled at regular intervals for a limited time period. When the signal is analyzed using discrete Fourier analysis, the analysis assumes that the data is periodic over the data collection period. This means that the data will be analyzed as if there were a periodic discontinuity in the signal where the two ends of the sample sequence are matched together. Such a discontinuity can create frequency components that were not present in the original signal. To reduce this effect, the sample data can be modified by multiplying the data sequence by a weighting or window sequence that reduces the amplitude of the signal at the end points of the sequence, minimizing the discontinuity.

The effects of extracting a time-limited portion of the signal and subsequent windowing can be analyzed mathematically.

Extracting a time-limited piece of a signal is equivalent mathematically to multiplying the signal by a square pulse or window. This window is zero outside the desired time segment, and unity within. Such a square window has a sin(x)/x behavior in the frequency domain as shown in Figure 2. Since the square time window is multiplied with the signal in the time domain, the signal's spectrum is convolved with the sin(x)/x frequency spectrum of the square window. The convolution operation results in replacing the single frequency line in the signal spectrum with the sin(x)/x function of the window, centered at the original frequency. The frequency corresponding to the reciprocal of the time duration of the window is the Fourier frequency. The discrete Fourier transform generate values from this continuous spectrum at discrete frequencies: integer multiples of the Fourier frequency. If the signal frequency falls exactly at an analysis frequency, then the zeros in the sin(x)/x function fall precisely at the analysis frequencies and only a single frequency component is observed. However, if the frequency of the signal falls between analysis frequencies, the analysis will show components at frequencies far removed from the original signal. This result is "smearing" of the original signal.

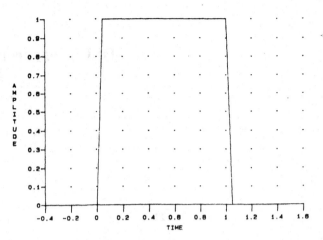

Figure 2a: Square Window

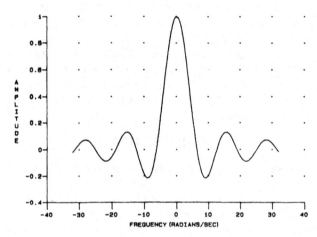

Figure 2b: Frequency Spectrum of Square window

By modifying the window, to taper the signal at the beginning and end of the signal, this smearing can be altered. Consider a general window that consists of the original square window multiplied by cosine waves that are harmonically related to the period of the square window. If the frequency spectrum of this window can be determined, then the smearing that would result from its use can be predicted.

A window consisting of cosine components multiplied by a square window can be analyzed if the cosine components are viewed temporarily as a signal. The frequency spectrum of a sum of cosine components, is a line spectrum. The height of each line component will be proportional to the amplitude of the corresponding cosine wave. When the sum of cosines is multiplied by the square window, each line in the spectrum is replaced by the sin(x)/x spectrum of the square window. The resulting spectrum can be viewed as a sum of sin(x)/x functions shifted in frequency as shown in Figure 3.

At first glance, analyzing such a sum appears to be very difficult. However, since the different cosine components of the window are harmonically related to the Fourier frequency, all the sin(x) terms in the sum are the same sin(x) function multiplied by either plus or minus one.

For example, a simple window could be written as follows:

$$h(t) = 1 + a^* \cos(2\pi t/T) \quad (-T/2 < t < T/2) \qquad (1)$$
$$0 \qquad \text{(otherwise)}$$

The corresponding frequency spectrum would be:

$$H(f) = \frac{.5a \sin\left(\pi T(f - \frac{1}{T})\right)}{\left(\pi T(f - \frac{1}{T})\right)} + \frac{\sin\left(\pi Tf\right)}{\left(\pi Tf\right)} +$$

$$\frac{.5a \sin\left(\pi T(f + \frac{1}{T})\right)}{\left(\pi T(f + \frac{1}{T})\right)} \qquad (2)$$

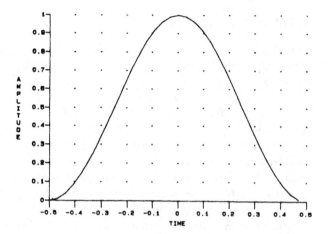

Figure 3a: Raised-Cosine (Hann) Window

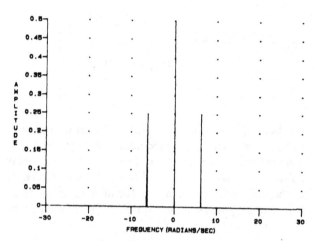

Figure 3b: Line spectrum of Hann Window

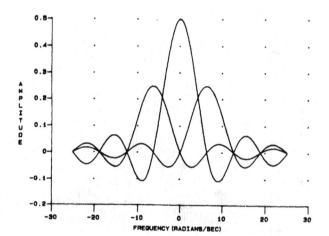

Figure 3c: Superimposed SIN(x)/x functions

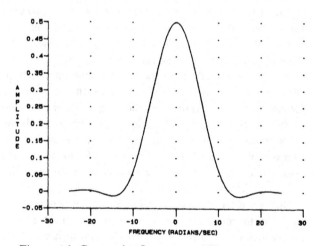

Figure 3d: Composite Spectrum of Hann Window

Since:

$$\sin(\pi(f \pm 1/T)^*T) = -\sin(\pi fT) \qquad (3)$$

The sin(x) from each of the terms can be factored out of the sum. The resulting expression consists of sin(x) multiplied by an expression:

$$H(f) = \frac{\sin(\pi Tf)}{\pi T} * \left(\frac{.5a}{(f - \frac{1}{T})} + \frac{1}{f} + \frac{.5a}{(f + \frac{1}{T})} \right) \qquad (4)$$

or:

$$H(f) = \frac{\sin(\pi Tf)}{\pi T} * \left(\frac{(1-a) f^2 + \frac{1}{T}2}{(f - \frac{1}{T})(f)(f + \frac{1}{T})} \right) \qquad (5)$$

As shown above in the simple case, and also in general when more cosine components are present, the denominator consists of a product of first order polynomials that are zero at the frequencies of each of the original cosine components. The numerator has zeros that can be positioned according to the proportion of cosine components present.

At each frequency where one of the products in the denominator goes to zero, there is also a zero in the sin(x) function. This means that the overall window spectrum will be tending to zero divided by zero at these frequencies. Since the spectrum of the window can also be expressed as a sum of sin(x)/x functions, the amplitude at these singular frequencies can be determined from the summation. As more cosine components are added to the window, the spectrum of the window becomes wider. However, with the increased width comes the option of choosing components that result in favorable characteristics in the numerator. If the proportion of cosine components is chosen so that the zeros of the numerator are at very high frequencies, then the window spectrum will be smaller at frequencies far removed from the center of the window spectrum. However, if as in the case of the Rosenfeld window, it is desired to minimize the window spectrum at nearby frequencies, one of the zeros can be placed just outside the primary lobe of the window spectrum.

The Rosenfeld window is designed using two cosine components, with frequencies equal to the Fourier

frequency and twice the Fourier frequency together with a constant or DC component. The width of the primary lobe of this window's spectrum is determined by the highest cosine frequency present in the window. Thus the window frequency width is plus or minus twice the Fourier frequency.

The proportion of the components in the Rosenfeld window are chosen so that the numerator has a pair of zeros at eight times the Fourier frequency. This results in excellent separation of signal components that are separated by at least eight times the Fourier frequency. The graph in Figure 4 shows the behavior of the Rosenfeld window compared with the more common Hann window.

MEASURING FREQUENCY

A method was sought for finding the frequency of signal components that were not mutliples of the Fourier frequency. Initial efforts were focused on developing the DSP equivalent of a "ratio detector" frequency discriminator found in broadcast FM receivers. Initial expectations were that such a discriminator would not possess the linearity required for instrument-grade performance. To achieve the desired level of performance, an interpolation algorithim was used to linearize the first experimental discriminators. However, it was found that by proper choice of discriminator coefficients, the linearity of the discriminator could be improved to the point where no correction was required within the computational limits of the processor used to test the technique.

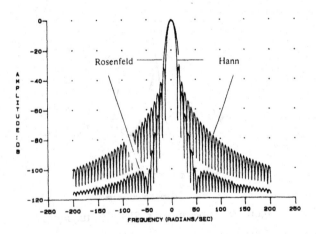

Figure 4: Rosenfeld and Hann Windows

Thus it was found that if the signal is windowed with a Rosenfeld window, a simple normalized, weighted sum of the components near the signal frequency yields the exact signal frequency. Furthermore, the weighting coefficients are independent of signal frequency, and of the number of data samples collected.

PERFORMANCE UNDER NON-IDEAL CONDITIONS

The technique was simulated with various forms of interference. In the presence of wide-band noise with a signal-to-noise ratio of 20dB, the peak observed error in frequency estimate was approximately 1/50 of the Fourier frequency and the average error was approximately 1/150 of the Fourier frequency. With a signal-to-noise ratio of 60dB, the peak observed error is about 1/6000 of the Fourier frequency and the average error is about 1/16000 of the Fourier frequency. These results are about four times better than could be expected using time measurement of the zero crossings of a sine-wave.

The effects of strong interfering components at other frequencies was also examined. The following table summarizes the findings when an interfering signal at half the level of the component of interest was present at the specified separation. The separation and errors are given in multiples of the Fourier frequency:

Separation	Average error	Peak error
1	.17	.36
3	.067	.094
4	.0093	.0147
5	.00091	.0014
6	.00019	.00031
7	.00005	.00008
8	.00002	.00003
9	.00002	.00003
10	.00001	.00003
11	.00002	.00003
12	.00002	.00003
13	.00002	.00003

Note that any DC offsets present in the signal will appear as a zero frequency component in the discrete Fourier analysis. This should be viewed as an interfering component. As such the lowest frequency that can be measured with small error is about six to eight times the Fourier frequency. This also insures that the second harmonic of the signal is at least six to eight times the Fourier frequency away from the fundamental frequency.

THE TECHNIQUE

The technique for applying this method is in two parts. The first part consists of generating the Rosenfeld window. This can be done during the first execution of the test program and need not be repeated on successive executions. The window can be generated from the following equations:

$$W(i) = (.762 - COS(2\pi \cdot i/N) + .238 \cdot COS(4\pi \cdot i/N))/A$$
where: $A = 1.05307$

This equation defines values of the window function for $i = 0$ to $N-1$ where N is the length of the window.

The value of A in the above equation was chosen for convenient computation of the amplitude of the signal component.

Stated using LTX-BASIC this would be implemented by the following statements:

```
10010  MAT W(1 TO M)=ZER \ REM SET WINDOW TO ZERO
10020   W(1)=.762/1.05307 \ REM SET DC COMPONENT
10030   W(3)=-1/1.05307 \ REM SET FIRST COSINE
10040   W(5)=.238/1.05307 \ REM SET SECOND COSINE
10050  MAT W(1 TO M)=INVERSE FFT(W(1 TO M)) \ REM
           CONVERT TO TIME
```

To determine the frequency of a signal component, multiply the collected data sequence by the Rosenfeld window. Next compute the magnitude of the frequency components at multiples of the Fourier frequency. This can be done by applying the FFT to the windowed data sequence, and then computing the RMS of the sine and cosine components at each frequency. Once the magnitude spectrum is computed, locate the frequency component within the expected range that has the largest value. If the magnitude spectrum is stored as a sequence named D, and the largest amplitude component is stored in $D(j)$, then the frequency of the signal is:

$$F = Ff \cdot \{ j + [D(j+1) - D(j-1) - .475 \cdot (D(j+2) - D(j-2))] \cdot 1.22453/D(j) \}$$

Where Ff is the Fourier frequency. The numerical constants in this expression were chosen to minimize the error in the frequency estimate. Figure 5 shows the departure from ideal for this frequency estimate.

The expression for determining frequency was chosen so that the frequency estimate would have the minimum error. The form of the equation was based on informed judgement and the coefficients were chosen by a least mean–squared–error regression to minimize error.

Once the largest element in the Fourier transform is located, the amplitude of the signal component can be accurately computed. The RMS of the seven spectral components from three times the Fourier frequency below the central component to three times the Fourier frequency above the central component will accurately measure the amplitude of the signal. If the Fourier transform algorithm is scaled so that the result gives the coefficient in a sine series, then the RMS will yield the PEAK value of the corresponding signal sine–wave component.

Stated using LTX–BASIC the technique is written as follows:

```
20010  MAT D(1 TO M)=D(1 TO M) * W(1 TO M)
20020  MAT D(1 TO M/2+1)=MAG(FFT(D(1 TO M))) QUIET
20030  MAT D(1 TO 4)=XTRM(D(F3 TO F4))
20040  J=D(2)+F3-2
20050     F=J+((D(J+2)-D(J))-.475*(D(J+3)-D(J-1)))
          /D(J+1)*1.22453
```

Where D( ) is the collected data sequence, W( ) is the Rosenfeld window sequence, F3 is the index of the lowest bin that could be the peak frequency, and F4 is the index of highest bin that could be the peak frequency.

Once the MAG(FFT( )) has been computed, the amplitude of the signal component can also be computed. The amplitude A is given by:

```
20060  MAT D(3 TO 3)=SSQ(D(J-2 TO J+4))
20070  A=D(3)^.5
```

In simulation, the maximum error in estimating amplitude, with various signal phases and frequencies, was approximately three parts per million.

A WORD OF CAUTION

The frequency measurement technique described above is very similar to the frequency detector used in an FM receiver. Like an FM receiver, if no signal is present, it will generate random results. Therefore, when this technique is used, the signal amplitude should always be checked to make certain that the expected signal is actually present.

CONCLUSION

A new DSP based technique for measurement of both frequency and amplitude has been presented. This technique has been tested extensively using both simulations and actual tests. In practice it has proven to be a useful technique for testing tone generators such as DTMF (dual tone, multi– frequency) telephone dialer circuits. With a single measurement, both amplitude and frequency of each signal component can be measured without special hardware filters.

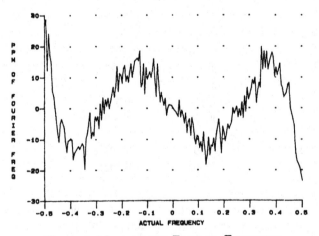

Figure 5: Measurement Error vs. Frequency

DSP SYNTHESIZED SIGNAL SOURCE
FOR ANALOG TESTING STIMULUS
AND NEW TEST METHOD

H. Kitayoshi, S. Sumida, K. Shirakawa, S. Takeshita

Advantest Corporation
(former Takeda Riken Co., Ltd.)
R & D and Engineering Center
1-16-1 Fujimi, Gyoda, Saitama 361, Japan

ABSTRACT

A new testing method, associated with a new signal source optimized for FFT (Fast Fourier Transform) analysis, has realized the transfer function of an analog DUT (Device Under Test) in an audio frequency band (0 - 100 kHz) in a wide dynamic range of 120 dB and with high precision of ± 0.002 dB and ± 0.02 degree and also estimated transfer function at 400 points at high speed ranging from several tens of msec to several hundreds of msec.

INTRODUCTION

This paper discusses a new signal source that permits high-speed, high-precision measurement of transfer function in a wide dynamic range, with reference to a theoretical basis for the new measurement process and means of its implementation.

The new method of transfer function estimation introduced herein consists of a signal source and a measuring technology that allow for the behavior and the effects of interiors of the DUT when compared with the heretofore available signal source and measuring technology [1] [2] [3]. This paper begins with discussions of the measuring precision of the method of transfer function measurement, and the factors affecting the measuring speed and the method of removing these effects.

a) Effects of DUT internal noises and their rejection
b) Effects of DUT network nonlinearity and their rejection
c) Effects of DUT transient responses and their rejection
d) Effects of A to D converters performance in measuring system

The paper then proceeds to introduce the new method of transfer function measurement featuring a system architecture and an autocalibration scheme which implement signal synthesis and signal processing in pipeline by using a DSP engine and a signal source hardware which generates specially synthesized signals in a Fourier space. The following four conditions are required for the stimulus synthesized by the DSP engine.

e) Small crest factor of the synthesized signal
f) Constant crest factor maintained within the DUT
g) Randomization of phase relations in the synthesized spectra series
h) Synthesizes source signal having a discrete spectrum compatible with the FFT algorithm

The paper concludes by presenting analysis of the following signals meeting the conditions enumerated in e) through h) above, their features, and results of experimentation:

i) T swept-sine wave (Time domain swept)
ii) F swept-sine wave (Frequency domain swept)
iii) Multi-swept-sine wave

METHOD OF TRANSFER FUNCTION
MEASUREMENT BY FFT AND ITS PRECISION

Figure 1 shows the configuration of a commonly used system for measuring the transfer function of an analog DUT on the principles of FFT technique. The signal applied to the DUT from the signal source and the signal output from the DUT are converted from analog to digital form, and transformed by FFT into frequency spectra $S_x(f)$ and $S_y(f)$ in each unit time-series frame. To eliminate the measurement error caused by the internal noise within the DUT and network nonlinearity, DUT input signal frequency power spectra $|S_x(f)|^2$ and DUT input/output signal frequency cross spectra $S_{xy}(f)$ are averaged in each measuring frame unit to obtain at the DUT transfer function $H(f)$ by solving:

$$H(f) = \frac{< S_{xy}(f) >}{< |S_x(f)|^2 >}$$

In the transfer function estimation system, higher processing speed could be achieved by averaging less frequently and estimating more spectra; on the other hand, higher precision result could be approached by averaging more frequently, increasing the evaluation spectra power, and protecting the DUT against distortion.

Hence, higher speed and higher precision are a matter of tradeoff.

Reprinted from *1985 International Test Conference*, 1985, pages 825-834.
Copyright © 1985 by The Institute of Electrical and Electronics Engineers, Inc.

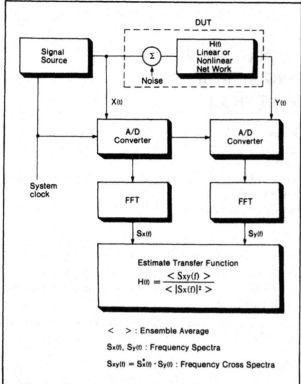

Sx(f), Sy(f) : Frequency Spectra

Sxy(f) = Sx*(f) · Sy(f) : Frequency Cross Spectra

Fig. 1   Transfer Function Measurement System Block Diagram Using FFT

Higher speed and higher precision in transfer function measurement are inhibited by the aforementioned factors a) through d), which are the topics for discussion in the following sections.

1-1. Effect of DUT Internal Noise

A signal source having minimum crest factor is required to eliminate the internally generated noise of the DUT for the high-precision, high-speed measurement of transfer function.

The cross spectra as observed is a vector sum of the DUT noise response vector $(S_x^*(f) \cdot Sn(f) \cdot H(f))$ and DUT input signal response vector $(S_x^*(f) \cdot Sx(f) \cdot H(f))$.

The noise response vector $(S_x^*(f) \cdot Sn(f) \cdot H(f))$ shows a random rotation from one measuring frame to another because there is no correlation between DUT input signal spectra $Sx(f)$ and noise signal spectra $Sn(f)$. As a result, the DUT transfer function measured after N times of frame averaging is expressed by

$$H(f) \cdot (1 + \varepsilon_h) = \frac{< Sxy(f) >}{< |Sx(f)|^2 >}$$

$$= \frac{S_x^*(f) \cdot Sx(f) \cdot H(f) + \dfrac{S_x^*(f) \cdot Sn(f) \cdot H(f)}{\sqrt{N}}}{S_x^*(f) \cdot Sx(f)}$$

$$= H(f) \cdot (1 + \frac{Sn(f)}{\sqrt{N} \, Sx(f)}) \qquad (*: \text{Conjugate})$$

The transfer function measurement error rate is given by

$$\varepsilon_h = \frac{Sn(f)}{\sqrt{N} \, Sx(f)}$$

High-precision estimation of the transfer function for a DUT having an internal noise source requires either increasing the number of times of frame averaging N or increasing the energy of input signal spectra $Sx(f)$.

1-2.   Effect of DUT Network Nonlinearity

The transfer function measurement error under the influence of DUT network nonlinearity can be reduced only by averaging of the frequency spectra series stimulated by a signal source having their phase relationships at random in each measuring frame. High-speed transfer function estimation can be achieved only by the use of a signal source having sufficiently small crest factor within the DUT.

DUT having nonlinear transmission characteristics. Second order harmonic distortion is assumed as DUT network nonlinearity.

$S_{xy}(f)$ is a vector sum of the DUT nonlinear response vector $(S_x^*(f) \cdot Sx(f/2) \cdot C(f))$ and the DUT linear response vector $(S_x^*(f) \cdot Sx(f) \cdot H(f))$.

Hence, if there is no correlation between $Sx(f)$ and $Sx(f/2)$ from one measuring frame to another, the error can be reduced by averaging in the same way as the internal DUT noise can be decreased.

1-3.   Effects of FFT Leakage Error and DUT Transient Responses

In the measurement of DUT transfer function by FFT processing, only an FFT analysis line spectra series should be provided and the analysis time series should be sampled only after the synchronous function synthesized by IFFT has been applied to the DUT and its transient response converged. The DUT output response to the input signal repeated by the periodic function is given by convolution integral of DUT impulse response function $h(\tau)$ and DUT input signal $x(t)$.

Since DUT output signal $y(t)$ observed immediately after the application of input signals to the DUT contains transient components of the DUT response, m·to sec. periodic function, a time-series truncation is generated from FFT processing, and impairing the precision of the transfer function estimation due to a spectrum leakage error.

DUT output signal $y(t)$ observed subsequent to time T sec. at which DUT impulse response $h(\tau)$ fully converges to zero forms an m·to sec. periodic function, having no leakage error from FFT processing.

1-4. Precision of Transfer Function Measurement in Infinite Bit A to D Converter and Dynamic Range

The greatest benefit of the method of transfer function measurement using FFT technique is that analysis of m/2 frequency spectra are made possible by m sampling points of unit frame data.

However, if more number of synthesized signal spectra are applied to the DUT, the available frequency spectra energy per frequency spectrum at a given signal amplitude will be decreased.

Hence, the observed spectrum is prone to the error effects of the internal DUT noise and the quantizing noise due to A to D conversion.

Figure 2 plots spectra power S_P per spectrum with regard to crest factors of 1.4 and 2.8 in the number of synthesized spectra L in multi-sine waves synthesized at a unit amplitude.

In other words, this figure represents the ability of the signal source in transfer function measurement to reject internal DUT noises effect.

Since the quantizing noise energy in the K bit A to D converter is constant regardless of the number of synthesized frequency spectra L, the S/N ratio for the observed spectra is given by

$$SN = 6K + 4.8 + S_P \text{ [dB]}$$

And the power accuracy of the observed spectra is

$$\varepsilon_\ell = 20 \cdot \log_{10} (1 + 10^{-\frac{SN}{20}}) \quad \text{[dB]}$$

where $\quad S_P = -10 \cdot \log_{10} (L) - 20 \cdot \log_{10} (Cf) \quad \text{[dB]}$

Cf: Crest Factor of Synthesized Signal

Figure 3 plots the spectra measurement accuracy with regard to number of bits K of an A to D conversion and number of synthesized spectra L.

Accordingly, this figure represents a measure of the transfer function measurement accuracy in the infinite bit A to D converter.

As for the phase of the observed spectra, which are calculated from the ratio of their imaginary components to the real components, the measurement accuracy is given by

$$\varepsilon_\theta = \frac{180}{\pi} \cdot 10^{-\frac{SN}{20}} \simeq 6 \cdot \varepsilon_\ell \quad \text{[deg.]}$$

The quantizing noise in A to D conversion is uniformly distributed across m/2 frequency spectra in the frequency domain, with the spectra power per spectrum being given by m/2.

Hence, the measuring dynamic range of the multi-sine wave is expressed by

$$Dr = SN + 10 \cdot \log_{10} (m/2) \quad \text{[dB]}$$

$$= 6 \cdot K + 4.8 - 10 \cdot \log_{10} (L) - 20 \cdot \log_{10} (Cf)$$
$$+ 10 \cdot \log_{10} (m/2) \quad \text{[dB]}$$

OPTIMAL SIGNAL SOURCE AND ITS CHARACTERISTICS

This section defines what an optimal signal source is that is suited to the measurement of transfer functions by FFT processing, and discusses its characteristics. Then, the following three new signal sources are proposed:

i)   T swept-sine wave
ii) F swept-sine wave
iii) Multi-swept-sine wave

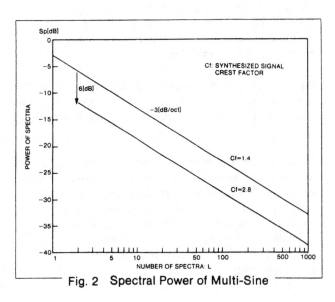

Fig. 2   Spectral Power of Multi-Sine

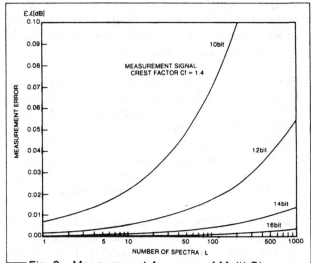

Fig. 3   Measurement Accuracy of Multi-Sine

These new functions, having something in common with one another, denote the progress of continual improvement being made on the measurement of transfer functions.

2-1. T Swept-Sine Wave

Authors will define a function as a signal source for use in the measurement of transfer functions by FFT processing, which permits a spectra series synthesized by required frequency power spectrum in the time domain.

In the m/2 frequency power spectra transformed by Fourier series for discrete time-series at m points, a time function that provides a flat frequency power spectra series from the l_1 spectra to the l_2 spectra is given by a sine wave frequency swept in a time domain, where time function $X(t)$ defined as follows:

$$\begin{cases} \omega\text{start} = 2\pi \cdot l_1 \\ \omega\text{stop} = 2\pi \cdot l_2 \end{cases}$$

where $\quad X(t) = \text{Sin}\{(\omega\text{start} + \omega(t)) \cdot t/m\}$

$$t: 0 \text{ through } m-1$$

where $\quad \omega(t) = \dfrac{t}{2 \cdot m} \cdot (\omega\text{stop} - \omega\text{start})$

Time function $X(t)$ continuously varies in frequency by $\Delta\omega/2$ from ωstart to ωstart + $\Delta\omega/2$ for the duration of t = 0 to m.

where

$$\begin{cases} \text{Sin}\{(\omega\text{start}) \cdot t/m\} & \text{When } X(0) \\ \text{Sin}\{(\omega\text{start} + \Delta\omega/2) \cdot t/m\} & \text{When } X(m) \end{cases}$$

where $\quad \Delta\omega = (\omega\text{stop} - \omega\text{start})$

Further, since $\omega(t)$ itself is a time function and involves a change of $\Delta\omega/2$ for the duration of t = 0 to m, $X(t)$ is subjected to phase modulation by $\omega(t)$, possessing a continuous power spectra series ranging from ωstart to ωstart + $\Delta\omega$ or from ωstart to ωstop.

Figure 4 shows the power spectrum of a synthesized time function that provides a continuous, flat frequency power spectra series from the 512nd spectra to the 1,536th spectra as synthesized in 4,096 point discrete time-series.

Because the T swept-sine wave shown here is caused by a sine wave frequency swept in the time domain, the amplitude histogram of the synthesized signal is identical to that of the sine wave, with a creset factor of about 1.4.

Thus, the T swept-sine wave synthesized in the time domain, with an extremely small crest factor and frequency band control, is found to provide satisfactory performance as a signal source for use in transfer function

measurement, when compared with signal sources heretofore available.

However, some possibility exists for further improvement in power spectrum flatness and spectrum power frequency band control as shown in Figure 4.

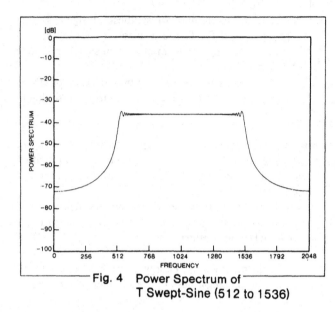

Fig. 4   Power Spectrum of T Swept-Sine (512 to 1536)

2-2. F Swept-Sine Wave

A time function offering good power spectrum flatness, complete spectrum power frequency band control, and an extremely small crest factor is synthesized in the frequency domain.

An optimized time function with complete power spectrum control and an extremely small crest factor could be obtained by using only the frequency spectrum information on the T swept-sine wave synthesized with a crest factor of about 1.4 and synthesizing the frequency spectrum in the frequency domain.

Figure 5 shows the frequency phase spectrum of the T swept-sine wave having the power spectrum shown in Figure 4. This frequency phase spectrum is apparently random.

Figure 6 plots the phase difference in the adjacent spectra of the phase spectrum in Figure 5. It shows a linear change from 0 to 360 degrees for the duration from ωstart to ωstop; that is, the differential phase in the frequency domain of the sine wave frequency swept signal in the time domain has an essential characteristics of varying linearly from 0 to 360 degrees within the synthesized power spectrum.

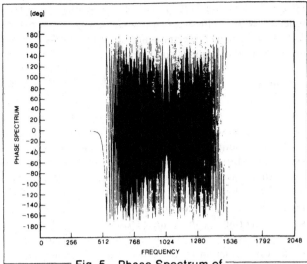

Fig. 5   Phase Spectrum of T Swept-Sine (512 to 1536)

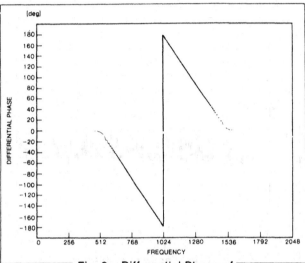

Fig. 6   Differential Phase of T Swept-Sine (512 to 1536)

Namely, time function $X(t)$ given by

$$X(t) = \sum_{k=l_1}^{l_2} Cos(2\pi \cdot k \cdot t/m + \theta_k)$$

where

$$\theta_k = \theta_{k-1} + (k - l_1) \cdot \theta_{inc}$$

where

$$\left| \begin{array}{l} \theta_{inc} = -2\pi/(l_2 - l_1) \\ \theta_{l_1-1} = 0 \end{array} \right.$$

is a synthesized function having only a flat power spectra series from the l_1 spectra to the l_2 spectra.

2-3.   DUT Response to Synthesized Signals and Their Optimization

Multi-swept-sine waves truly optimized for the measurement of analog DUT transfer functions have been synthe-

sized by defining and correcting drawbacks of analog DUT transfer function measurement based on swept-sine waves.

A multi-sine wave (F swept-sine wave) synthesized in the frequency domain, having a crest factor of 1.5 through 1.8, is no doubt a nearly optimal function as a signal source for use in transfer function measurement.

It should be noticed here, however, that, even the crest factor of the DUT input signal may be low enough, better transfer function measurement precision and a wider dynamic range cannot be achieved unless the crest factor of the DUT output signal is low. This means that the transfer function measurement precision and the dynamic range can be improved not only by lowering the crest factor of the DUT input signal but also lowering the crest factor of the DUT internal signal to prevent DUT distortion on the same energy and that of the DUT output signal to improve the A to D conversion precision and dynamic range. Figure 7 shows maximum crest factor Cmf of the signal wave that is observed by applying frequency band select Sw to the frequency spectrum of an F swept-sine wave to estimate the effects of a frequency spectrum band select by the DUT.

Figure 7 shows computer simulations of partial spectrum width Sw versus maximum crest factor Cmf for F swept-sine waves synthesized with number of frequency spectra $L = 20, 40, 80, 160, 320,$ and 640 with respect to 4,096 point time data. These curves are obtained by locating frequency band spectra 50 times randomly with a spectrum frequency band select and plotting the maximum crest factor among them with respect to the bandwidth. The simulation results in Figure 7 indicate that the crest factor of the swept-sine wave is increased significantly by a spec-

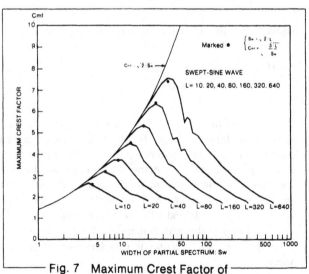

Fig. 7   Maximum Crest Factor of Partial Spectrum (F Swept-Sine)

trum frequency band select of the signal. The maximum crest factor associated with the frequency band select of $Sw = \sqrt{2 \cdot L}$ is predicted to be $Cmf = \sqrt{3 \cdot \sqrt{L} / \sqrt{2}}$
For example, the F swept-sine wave synthesized by number of frequency spectra L = 320 is subjected to a frequency band select of 25 spectra, the crest factor of the signal rises to about 6.3. This indicates that the swept-sine wave synthesized from a series of L spectra has time-series energy concentration in the frequency band neighboring spectrum of $\sqrt{2 \cdot L}$. Thus, the measurement of DUT transfer functions based on swept-sine waves easily distorts the DUT output signal and reduces the DUT output signal A to D conversion precision and dynamic range.

The curve $Cmf = \sqrt{2 \cdot Sw}$ shown in Figure 7 denotes the worst crest factor of the partial spectrum for impulse waves.

For the reasons stated above, the method of measuring DUT transfer functions by swept-sine waves should be used only where the DUT is subject to very little frequency band select or where only a few synthesized spectra (up to 20) are involved.

The signal source required in practical transfer function measurement should have an extremely small crest factor of the input signal itself, which is virtually unaffected by an anticipated frequency band select of the frequency spectrum within the DUT.

Then, the authors attempted a dispersion of the time-series energy concentration of a narrow band neighboring spectrum in an F swept-sine wave, which is expressed by

$$X(t) = \sum_{k=l_1}^{l_2} Cos(2\pi \cdot k \cdot t/m + \theta k)$$

where $\qquad \theta k = \theta k-1 + (k - l_1) \cdot \theta inc$

where $\qquad \begin{vmatrix} \theta inc = -2\pi \cdot \alpha / (l_2 - l_1) \\ \theta_{l_1-1} = 0 \end{vmatrix}$

Here, new coefficient α for θinc is called a swept phase acceleration factor. Time function $X(t)$ synthesized possesses only a flat frequency power spectra series from the l_1 spectra to the l_2 spectra in the frequency domain. The new coefficient is to adjust the phase relation of the spectra series. Namely, $X(t)$ denotes an F swept-sine wave when $\alpha = 1.0$, and when $\alpha > 1.0$, frequency sweep exists α times in the time domain for the duration of t = 0 to m. The new function is called a multi-swept-sine wave.

Figure 8 plots the crest factor of the synthesized time function with multi-swept-sine wave having a series of 400 identical power spectra synthesized from 4,096 point time data in the frequency domain.

As can be seen from Figure 8, the swept-sine wave with $\alpha = 1.0$ shows the lowest crest factor of about 1.7. Even when $\alpha > 1.0$, the value of α that provides a crest factor of 2.0 or less is selectable.

As for the method of determining the value of α in the practical multi-swept-sine wave, because the multi-swept-sine wave is generated by repeating a swept-sine wave for each frequency band, the unit swept-sine wave involved in a synthesized series of L spectra is L/α spectra series. Accordingly, the value of L/α must be 20 or less. If it exceeds 20, the partial spectrum crest factor in its synthesized function is always larger than 3.0 when $Sw = \sqrt{2 \cdot L/\alpha}$.

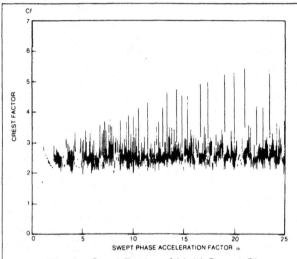

Fig. 8   Crest Factor of Multi-Swept-Sine (1 to 400)

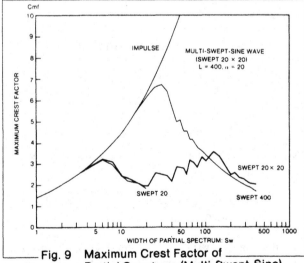

Fig. 9   Maximum Crest Factor of Partial Spectrum (Multi-Swept-Sine)

Figure 9 shows computer simulations of maximum crest factor Cmf with respect to partial spectrum width Sw of the multi-swept-sine wave (SWEPT 20 × 20) synthesized from 4,096 point time data with L = 400 and α = 20.

As can be seen from Figure 9, the maximum crest factor shows a value close to the F swept-sine wave of L = 20 (SWEPT 20) when partial spectrum width Sw is low, a value close to the F swept-sine wave of L = 400, (SWEPT 400) when Sw is high.

Figures 10 and 11 show the input and response waveforms and power spectra of F swept-sine wave (SWEPT 400) and multi-swept-sine wave (SWEPT 20 × 20) synthesizes from a series of 400 identical power frequency spectra for a DUT having spectrum frequency band select characteristics. In the figures, applied signals are shown in (a), DUT response signals in (b). A comparison of Figure 10 (b) with Figure 11 (b) indicates that the DUT

output signal for the swept-sine wave has a very large crest factor with much time-series energy concentration, while the DUT output signal for the multi-swept-sine wave has a relatively low crest factor with small time-series energy concentration.

The DUT used in this measurement was a telecommunications transfer filter (Intel 2912A). Measurement was made by the use of a 15-bit A to D converters.

The practicability of predicting the creset factor of a swept-sine wave with a frequency band select has so far been discussed, along with the the DUT output signal being easily distorted under stress during measurement of transfer functions based on swept-sine waves.

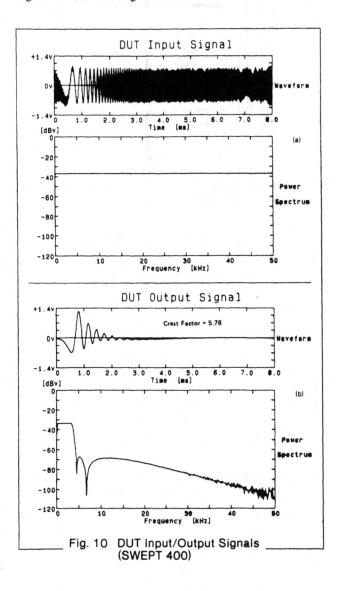

Fig. 10   DUT Input/Output Signals
(SWEPT 400)

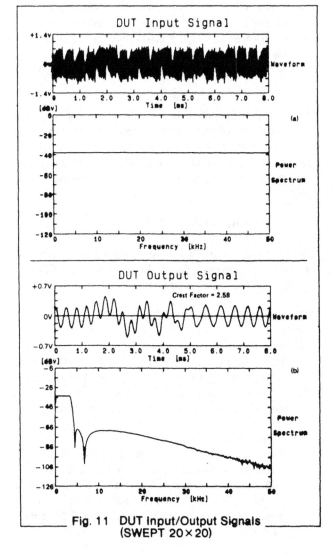

Fig. 11   DUT Input/Output Signals
(SWEPT 20 × 20)

Also, a multi-swept-sine wave has been induced as a new signal source that has a significantly small crest factor for the input signal itself, which is virtually unaffected by a frequency band select of the DUT. By appropriately selecting swept phase acceleration factor for each frame, a signal with a small crest factor can be generated which has a random phase relation from frame to frame in the synthesized spectra series. As stated in the section 1-2, this process of signal generation is essential particularly to the measurement of nonlinear network transfer functions.

NEW PROPOSALS FOR TRANSFER FUNCTION ESTIMATION IN HIGH SPEED, HIGH PRECISION, WITH WIDE DYNAMIC RANGE

On the basis of the discussions presented so far, the following three proposals are made for the estimation of high speed, high precision, and wide dynamic range transfer functions and are demonstrated:

i)   Advanced parallel processing for higher speed, and determination of the optimal measurement timing
ii) High speed and high precision system calibration for higher precision
iii) Spectrum block sweep to obtain a wider dynamic range

In this paper, the 400 point transfer function of an analog DUT is estimated by using these technologies with a gain accuracy of ± 0.002 dB and a phase accuracy of ± 0.02 degree in a dynamic range of 120 dB in an short estimation period on the order of several tens of msec to several hundred of msec.

3-1. System Architecture

Figure 12 shows the hardware configuration of the new measurement process. Figure 13 shows its data flow. Its configuration is reviewed in further detail as follows:

Hardware configuration (See Figure 12.)

(1) A to D converters
Incorporating an antialiasing filter, a 16K word buffer memory with a 15-bit precision A to D converter samples DUT input/output signals in 2.56 times sampling rate of the estimated frequency band.

(2) Signal source
Incorporating a 12 bit D to A converter, a time interpolation low pass filter, and two 4K word waveform buffer memories with a 10.24 times conversion rate of the estimated frequency band, the signal source generates functions synthesized by the DSP engine as analog signals.

(3) DSP engine
A 32-bit floating-point array processor, the DSP engine synthesizes applied signal functions and estimates transfer functions.

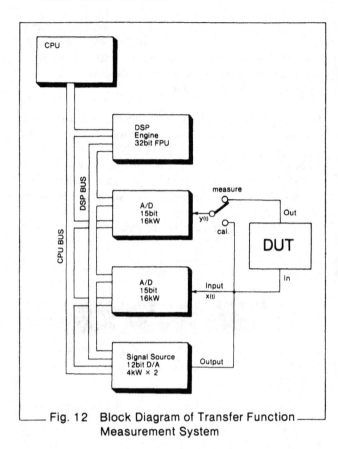

Fig. 12 Block Diagram of Transfer Function Measurement System

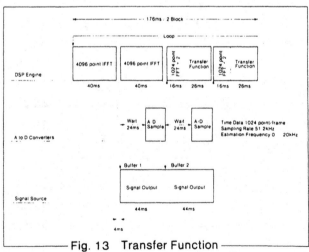

Fig. 13 Transfer Function Measurement Data Flow

Software functions (See Figure 13.)

(1) Higher speed
  i) The estimated frequency band is split and the number of synthesized spectra optimized from the predicted measurement precision given in Figures 2 and 3, so that the transfer function having a required precision can be measured in a minimum time.
  ii) Pipeline processing is performed to speed up signal synthesis, measurement, and transfer function estimation.
  iii) System data flow is simplified by synthesizing multi-swept-sine waves in any frequency band from three parameters (l_1, l_2, α) with a special signal synthesis and processing DSP engine.

(2) Higher precision
  i) Multi-swept sine waves are selectively used as shown in Figures 7, 8, and 9 to minimize the crest factors of the DUT internal and output signals for higher precision.
  ii) Added measuring precision is provided by use of high speed, high precision automatic system calibration to correct the transfer function measurement error resulting from the large difference between the input voltage ranges of two A to D converters [4].
  iii) The optimal suspending time from the moment of signal input to that of measurement is determined so as to eliminate the FFT leakage error with a choice of the measurement timing.

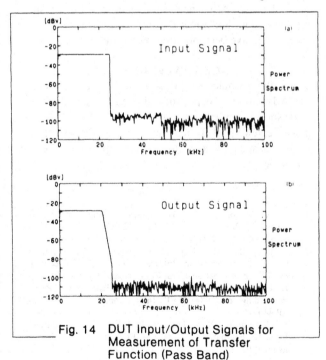

Fig. 14 DUT Input/Output Signals for Measurement of Transfer Function (Pass Band)

3-2. Practical Measurement of Audio PCM Filter Transfer Functions by Spectrum Block Sweep

It is mandatory with a PCM audio low-pass filter to measure the pass band gain to a precision of 0.01 dB and estimate the stop band attenuation in a 100 dB dynamic range.

The transfer functions for such a DUT must separately be estimated over the entire frequency band by measuring

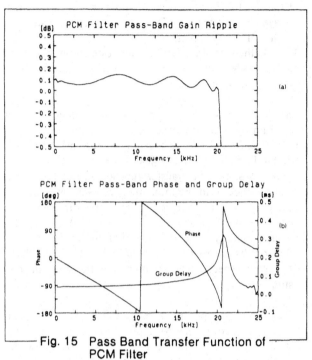

Fig. 15 Pass Band Transfer Function of PCM Filter

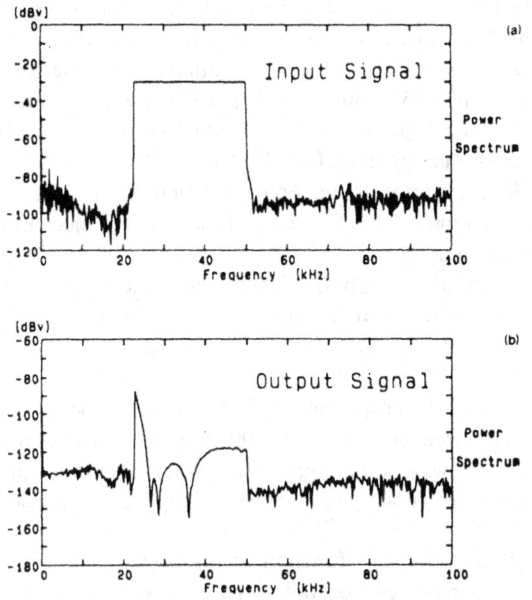

Fig. 16 DUT Input/Output Signals for Measurement of Transfer Function (Stop Band)

the pass band transfer function and the stop band transfer function; that is, multi-swept-sine waves controlled over the synthesized frequency power spectrum frequency band are used to measure transfer functions at the application of pass band and stop band signals. This makes the estimated spectra levels of DUT response output signals during transfer function measurement essentially equal to each other, thus permitting high-precision estimation.

This signifies a widened dynamic range for the estimation of transfer functions associated with a change in the input voltage range of the A to D converter from one estimated spectra series to another.

This method of transfer function estimation has been made possible by the combined use of high-precision system calibration [4].

Figure 14 shows the waveforms and frequency power spectrum of 100 frequency spectra synthesized signal (14a) applied to measure the pass band transfer function and DUT response output signal (14b).

Figure 15 shows the transfer function in the DUT pass band obtained from the measurement results presented in Figure 14. It is not averaged. Gain versus frequency response curve are given in (15a), phase and group delay versus frequency response curve in (15b). These measurement results are well indicative of the system's capability of higher speed, higher precision transfer function estimation.

Figure 16 shows the time domain waveform and frequency power spectrum of 110 frequency spectra synthesized signal (16a) applied to measure the stop band transfer function and DUT response output signal (16b).

In the measurement of Figure 16, the A to D conversion for DUT input signal is performed in the 0 dBV input voltage range and that for DUT output signal is performed in the −60 dBV input voltage range. The DUT output signal after the high-pass filter are subjected to an A to D conversion to eliminate DC offset from them.

Hence, a transfer function correction is carried out in the estimation of DUT transfer functions to reject the high-pass filter effects.

System's capability of higher speed, wider dynamic range transfer function estimation should be evident from a look at the measurement results in Figure 16.

Figure 17 shows the DUT transfer functions over the entire frequency band at 400 frequency points measured in this way. Gain versus frequency response curve are given in (17a), phase versus frequency response curve (17b).

In the transfer function measurement of Figure 17, averaging is carried out twice for the pass band transfer function in one block and four times for the stop band transfer function in two blocks. The time needed to esti-

mate transfer functions over the entire frequency band is as short as 440 msec.

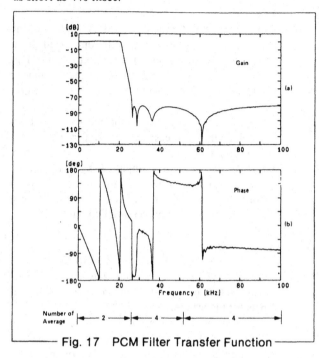

Fig. 17 PCM Filter Transfer Function

CONCLUSION

This paper has discussed a scheme of optimizing the measurement of transfer functions by FFT processing and also examined characteristics of the synthesized signal functions for the measurement of transfer functions.

As the result, an optimal method of signal synthesis for the measurement of analog DUT transfer functions has been derived. Also, new techniques for higher speed, high precision, and wider dynamic range transfer function estimation have been proposed and demonstrated.

ACKNOWLEDGEMENT

The authors would like to thank Dr. H. Sasaki, President of Advantest Corp. for his encouragement and suggestion. Also acknowledge the valuable assistance and paper reviewing of T. Kazamaki and S. Kamata. Thanks to T. Nakatani for assisting in the preparation of experimental system.

REFERENCES

[1] M. V. Mahoney "New Techniques for High Speed Analog Testing" Proceedings of 1983 IEEE International Test Conference, PP.589 - 597

[2] J. Hofer, B. Sigsby "Digital Signal Processing Test Techniques for Telecommunication Integrated Circuits" Proceeding of 1983 IEEE International Test Conference, PP.750 - 770

[3] J. F. Campbell "Transfer Function Estimation, Part II" Proceeding of 1984 IEEE International Test Conference, PP.440 - 446

[4] H. Kitayoshi "New Test Method and Signal Source for Efficient Estimation of Transfer Function" 1985 Advantest Internal Paper.

An NBS Calibration Service for A/D and D/A Converters

by T. M. Souders and D. R. Flach

Electrosystems Division
National Bureau of Standards
Washington, DC 20234

Abstract

An NBS calibration service for high performance 12- to 18-bit analog-to-digital and digital-to-analog converters (ADC's and DAC's) is described. The service offers comprehensive measurements of linearity, differential linearity, gain, offset, and rms input noise (for ADC's), with systematic uncertainties as low as 3 ppm. Measurements are made at a minimum of 1024 different codewords. The measurement approach, design features, test programs, and data reduction techniques are discussed, as are the methods of error estimation and quality control.

Introduction

In response to the increasing importance of data converters in modern measurement and control instrumentation, the National Bureau of Standards is offering a calibration service for measuring the transfer characteristics of high performance analog-to-digital and digital-to-analog converters. This service specifically addresses the measurement of static converter errors, however, a capability for making dynamic measurements is currently under development.

The service will be of particular importance to manufacturers and users of converters who want to: test converters destined for highly demanding applications, satisfy traceability requirements imposed by military or government contracts, obtain independent verification of their own test procedures through the use of transfer standards, verify the performance of converters used in automatic test equipment (as part of an ATE calibration program), or test experimental converter designs during development. Also, with the bit error correction data provided (as described later), improved linearity performance can be realized.

The principal error parameters measured are those of linearity and differential linearity. Nevertheless, gain and offset errors may also be measured when appropriate, and the equivalent rms input noise of ADC's can be measured as well. Determination of monotonicity in DAC's and missing codes in ADC's is not generally performed, since it requires excessive measurement time for high resolution converters. These characteristics can often be inferred, however, from the linearity data available.

Since data converter terminology has so far not been standardized, terms used in this publication are defined as follows:

1. Linearity Error: The difference between the actual and ideal levels of the static input/output characteristic after offset and gain errors have been removed.

2. Differential Linearity Error: The difference between the actual and ideal separation between adjacent levels.

3. Offset Error: The difference between the actual and ideal levels, measured at the most-negative level of the input/output characteristic.

4. Gain Error: The difference between the actual and ideal levels, measured at the most-positive level of the input/output characteristic, after the offset error has been removed.

5. Equivalent RMS Input Noise: The rms value of the effective internal noise of an ADC, referred to its input terminals.

In each of these definitions, the specified level is taken to be: for DAC's, the discrete (analog) output level corresponding to a particular codeword, and for ADC's, the analog input level at which the transition between adjacent codes occurs. In this later case, the upper digital code is taken to define the transition.

Data Converter Test Set

The service is based on an automated test set developed at NBS in which the converter under test is compared with a reference DAC [1]. A 20-bit plus sign, relay-switched converter also developed at NBS, is used for this purpose [2]. It exhibits less than 1 ppm linearity error, and incorporates an easily used self-calibration feature.

For testing DAC's, test codewords are input to both the reference and test converters and their outputs are compared directly (Fig. 1). Differences in outputs are amplified, digitized with 1 ppm resolution, and returned to the controller for processing.

ADC's are compared with the reference by placing them within a feedback loop [1,3] locking the input voltage to a code transition level defined by the input codeword to the reference converter as illustrated in figure 2. This input voltage is, in turn, compared to the output from the reference DAC. The precision, ΔV, with which transition levels are located is determined by the integration time constant, the conversion rate, and the integrator's input voltage. The conversion rate is normally fixed at 10 kHz, however, the integrator's input voltage is programmable, and is typically set to give a locking precision of 1/16 LSB.

The transition-locking feedback loop employed in this test set offers several advantages over other measurement approaches employing a DAC to directly excite the ADC under test. First, the transition is automatically located with high precision in the presence of significant noise in the test unit itself. Second, a wide-band very-low-noise output amplifier operating with essentially unity noise gain can be selected as the integrating amplifier which excites the ADC input. Furthermore, since the feedback loop locks onto the transition voltage of the test converter, drifts due to time or temperature in the feedback amplifier itself are not as critical as they would be in a reference DAC's output amplifier. Third, the feedback loop can be used to implement an algorithm facilitating an automatic measurement of the equivalent rms input noise of the ADC, as described next.

ADC RMS Noise Measurement

Using this technique [4], the noise measurements are made in the vicinity of the converter's decision levels represented by code transitions, where the noise sensitivity is greatest. These transition levels are located and locked onto with the feedback loop controlling the input voltage to the converter under test as previously described (see Fig. 2). The actual rms noise measurement is based on a theoretical relationship between input noise and an expected number of counts derived digitally from the response of the feedback circuit.

With a noiseless ADC in the feedback loop, the input voltage change, once locking has occurred, will reverse its slope after each conversion describing a triangular waveform with peak-to-peak amplitude equal to the locking precision ΔV, as in figure 3-a. (The magnitude ΔV is determined by clock period, integration time constant, and integrator input voltage, as previously noted. Its value is set from the front panel.) The addition of random noise of rms value σ at the ADC input causes a corresponding change in the feedback response, as shown in figure 3, b-d. The input voltage now follows a "random walk" about the transition level, reversing its slope with decreasing frequency (and straying farther from the transition level) as the noise level increases. If, at the given sampling rate, the successive values of noise are uncorrelated and follow a Gaussian distribution, the statistics of the random walk can be calculated in terms of the rms noise level. An easily measured statistical parameter has been selected which can form the basis for an equivalent noise measurement made in terms of the voltage ΔV. In particular, the probability P of occurrence of a slope reversal following a conversion is easily measured with digital circuitry, since slope reversal is controlled by the output of a digital comparator. This probability can be accurately related to noise level as described in reference 4. Since ΔV provides the reference voltage for the measurement of noise value σ, it is convenient to express P in terms of the ratio $\sigma/\Delta V$.

A plot of the ratio $\sigma/\Delta V$ vs the calculated reversal probability P is given in figure 4. A mathematical analysis of this technique is provided in reference [4], together with a discussion of error sources and experimental verification.

General Test Strategy

Since many high accuracy, high speed ADC's and almost all DAC's operate with intrinsically fixed values assigned to each individual binary digit, any output or input codeword (comprised of a linear combination of these bits) has an assigned value theoretically equal to the sum of the values of the individual bits that are included. Were it not for small interactions between the bits and other code-dependent effects collectively called superposition errors, n measurements would suffice to completely characterize the 2^n code states of an n-bit converter. Because significant superposition errors frequently do exist, the NBS test facility is equipped to measure many, if not all, of the 2^n states.

Typically, all 1024 digital codeword combinations of the 10 most significant bits are measured. Errors contributed by the remaining less significant bits are generally insignificant, a premise which is tested during the calibration process. From the test data are calculated the offset and gain errors and the linearity errors associated with each tested codeword.

Basic Linearity Test

For the basic linearity test, the 1024 codewords formed from the 10 most significant bits are tested in numerically ascending order. If the unit-under-test is an ADC, these reference codewords designate the upper codes of digital transitions, so that the measured input voltage corresponds to the transition between the designated codeword and the adjacent codeword below it. Since no ADC transition exists for a reference code of all zeros, the next higher code (0000000000---1) is chosen instead for the most-negative test code. For simplicity, this convention is maintained for DAC's as well. At each code the test set automatically measures the errors, defined as the difference between the input (output) of the ADC (DAC) under test and the output of the reference DAC-20. Errors in DAC-20 are assumed to be negligible. These measured errors are numerically corrected for offset and gain, forcing the first and last code errors to zero. The resulting values ε_n are the linearity errors of the converter-under-test and are recorded. These data are numerically processed to determine the maximum, minimum, and rms errors of the 1024 tested codes, and to determine, on a least squares basis, individual correction coefficients for the 10 most significant bits.

Walsh Function Analysis

The correction coefficients just mentioned are derived from the Walsh function expansion of the test converter's error data [5]. In this analysis, the set of error data derived from the basic linearity test is taken to represent a uniformly sampled function having a period of 2^N samples, where N, in this case, is 10.

Walsh functions are a set of complete orthogonal functions capable of representing, by a simple series, essentially any such function on a finite interval [6]. Such a series is analogous to a Fourier series, differing primarily in that Walsh functions are square-like rather than sinusoidal. Three properties of Walsh functions make them particularly useful in analyzing converter errors:

1. Walsh functions are orthogonal, and as such, each function of the series independently minimizes the mean-squared error in the representation of the original function.

2. Walsh functions are square-like, and assume a value of plus or minus one.

3. The set of 2^N Walsh functions contains a subset of N Rademacher functions which are exact square waves having periods of 2^n samples, where n ranges from 1 to N. This relationship is illustrated in figure 5, for N=3.

By applying property 3, it can be seen that the Rademacher functions have the same periodicity as the individual bits in a linearly ascending digital ramp as shown in the figure. If the on and off states of a bit are defined as +1 and -1, respectively, then by property 2, for N=10, the 10 Rademacher functions exactly describe the digital states of the 10 bits tested in the 1024-point linearity test described above.

Finally, property 1 implies that the individual Rademacher coefficients will give the errors associated with each respective bit in the sense of the best (least-squares error) fit, independent of bit interaction or superposition error. Therefore, if no superposition error is present, a set of 10 Rademacher functions will completely describe the linearity errors of a 10-bit converter. Furthermore, if superposition errors are present, the 10 coefficients still describe the best corrections which can be applied to each individual bit when such corrections are considered over the entire range of code states. If such corrections based on the Rademacher coefficients were applied, the remaining (superposition) errors would have the minimum possible mean-square value based on any method of individual bit adjustments. In such cases where superposition errors are present, the remaining mean-square error can, of course, be further reduced by the inclusion of additional Walsh coefficients beyond the Rademacher series. Of particular interest is the 0th Walsh coefficient, which by analogy to the DC component in Fourier analysis provides the net offset or averge value of the data set. Note that since the linearity errors are by definition zero at the endpoints, an offset in the error data can occur only through the presence of superposition errors. While other Walsh coefficients could also be considered, usually only the 0th Walsh and the N Rademacher coefficients have an easily recognizable physical significance. And even for these coefficients, the significance is lost for certain converter types. For example, a flash-type ADC having 2^N-1 independent threshold circuits would be unlikely to exhibit significant correlation between linearity errors and the Rademacher series. Nevertheless, many ADC types, including successive approximation, and almost all DAC types will profit by this type of analysis.

The formulas used to calculate the rms error as well as the 0th Walsh and N (10) Rademacher coefficients are as follows:

$$rms\ error = \sqrt{\frac{1}{1024} \sum_{n=0}^{1023} (\epsilon_n)^2}$$

$$C_o = \frac{1}{1024} \sum_{n=0}^{1023} \epsilon_n$$

$$C_M = \frac{1}{1024} \sum_{n=0}^{1023} \epsilon_n (2K_{Mn}-1)$$

where ϵ_n is the linearity error of the nth code as defined previously

C_o is the 0th Walsh coefficient or the average of errors ϵ_n

C_M is the Mth Rademacher coefficient, or the correction coefficient for the Mth bit

K_{Mn} is the logic value of the Mth bit in the nth codeword. It has a value of either 1 or 0.

As the preceding discussion suggests, the determination of selected Walsh coefficients

can indicate the extent to which superposition errors are present.

To determine the superposition errors from the correction coefficients, the coefficients are transformed back to a function of the original form, reconstructing the data set. The differences between the measured and reconstructed data are then, by definition, the converter's superposition errors. The reconstruction is calculated with the following formula:

$$\epsilon_{Rn} = C_o + \sum_{M=1}^{10} C_M(2K_{Mn}-1)$$

where ϵ_{Rn} is the reconstructed error for the nth codeword.

Superposition errors are then defined as

$$\epsilon_{sn} = \epsilon_n - \epsilon_{Rn},$$

and a superposition "figure of merit" is given by the rms value of ϵ_{sn} computed over the set of n codewords:

$$rms = \sqrt{\frac{1}{1024} \sum_{n=0}^{1023} (\epsilon_{sn})^2}$$

This information can uncover inherent design problems which cause effective coupling between bits, for example. In general, superposition errors establish a practical limit to the accuracy attainable with a given converter, beyond which no simple adjustments or corrections will be useful. If, however, the superposition errors are significantly lower than the measured linearity errors, then substantial improvement in performance might be possible through simple application of the correction coefficients, either by direct adjustment or via system software. Figure 6 presents plots illustrating two converters having, respectively, (a) small superposition errors, and (b) relatively large superposition errors.

The basic linearity test program is capable of plotting all 1024 points each of error data, reconstructed error data, and superposition errors vs digital codeword . The maximum, minimum, and rms errors of these data sets are each determined, as are the eleven error coefficients.

Differential Linearity Test

By definition, differential linearity errors (DLE's) are the errors in separation between adjacent code levels. Consequently, DLE's are associated with the discontinuities observed in the plots of linearity error vs codeword: the greater the discontinuity, the larger the DLE at that location. Assuming as before that the bits less significant than the 10th contribute insignificant errors, then the differential linearity errors for a particular converter can be easily determined by inspection from the plots of linearity error. Since the major discontinuities almost always occur at major transitions, an abbreviated test can provide actual measurements of the DLE's at the codes for which the errors are likely to be greatest. The following sequence of 2 (N-1) codeword pairs is used for this test:

```
  MSB            LSB
  1 2 3. . . . . N

  0 0 0. . . 0 0 1⎫
  0 0 0. . . 0 1 0⎭
  0 0 0. . . 0 1 1⎫
  0 0 0. . . 1 0 0⎭
        .       .
        .       .
        .       .
  0 1 1. . . 1 1 1⎫
  1 0 0. . . 0 0 0⎭
  1 0 1. . . 1 1 1⎫
  1 1 0. . . 0 0 0⎭
        .       .
        .       .
        .       .
  1 1 1. . . 0 1 1⎫
  1 1 1. . . 1 0 0⎭
  1 1 1. . . 1 0 1⎫
  1 1 1. . . 1 1 0⎭
  1 1 1. . . 1 1 0⎫
  1 1 1. . . 1 1 1⎭
```

Differential linearity errors at each of the code pairs are simply computed as

$$\varepsilon_{DL} = \varepsilon_2 - \varepsilon_1$$

where ε_2 and ε_1 are the measured errors at the 2nd (higher) and 1st (lower) codes, respectively, of the code pair.

Random Code Test

While the test of linearity errors discussed previously is based on a large number of data points, practical constraints generally make it impossible to cover all possible codes, much less all code sequences. The potential complexity of these untested residual error sources suggests that they be estimated instead by statistical means and be included in the estimate of the random error associated with the reported data. For this purpose, a random code test has been developed and is applied to each converter-under-test.

In this test, test vectors are randomly selected from the full set of 2^N possible codewords, and 1024 of these are tested in succession. Each codeword has the full word length of the unit-under-test, so that the less significant bits are randomly included as well as the 10 most significant bits. The measured errors are again corrected for offset and gain, using the same endpoint test codes as are used in the basic linearity test. The resulting random code linearity errors are sorted into 1024 possible bins according to the first 10 bits of each codeword. In this manner, it is possible to relate each random codeword error to one of the 1024 ordered errors measured in the basic linearity test. Note that, since the codewords are randomly selected, not all 1024 bins will be filled; in fact, it can be shown that an average of approximately 377 bins will remain empty. When the set of errors measured in the basic linearity test is subtracted from the sorted set of random errors (ignoring bins that are not filled), then there remains a set of residual errors. These residuals are comprised of errors from several sources. First is the error contribution attributable to the less significant bits and their possible interactions with the more significant bits. Second are the effects due to a random test sequence as opposed to a well-ordered, linear sequence. Internal thermal effects might be prominent among these. Finally, there will be included other random errors of the measurement process which limit precision even when the same test sequence is repeated. Taken together, these effects give the residual errors a random appearance, and for simplicity, they are treated as being random. Accordingly, the rms value of the set of approximately 647 residual errors is calculated, and this value is taken to be an overall measure of the random error of the calibration process. This figure is used in determining the uncertainty limits that are stated in the calibration report.

ADC Noise Test

The algorithm used in making ADC noise measurements is based upon two assumptions: that the noise to be measured has a Gaussian distribution, and at the given sampling rate (10 kHz), successive values are uncorrelated. These assumptions (which usually prove to be valid), are easily tested by varying ΔV and performing a least-squares fit of noise data to the theoretical curve of figure 4.

Once the validity of the model has been demonstrated, either of two noise tests may be run: a determination of the average noise measured over 1024 random codes, or a test of noise vs code, measured and plotted at 64 randomly selected codes. In the average noise test, the 1024 reversal probabilities are averaged. From this and the ΔV setting, the average, randomly sampled noise is computed. If the variation of noise with output code is desired, the second test may be run. For this test, the number of reversals is measured 100 times at each of 64 randomly selected codewords. The noise is calculated for each code from the average of the 100 measurements, and is plotted.

Test Capabilities

The systematic uncertainties in measuring the various converter error parameters have been analyzed and their estimated limits are summarized in table 1. Random errors of the measurement process necessarily include random variations in the test converter and are therefore individually evaluated for each test as previously discussed. Nevertheless, measurements on very stable converters indicated the random errors contributed by the test set itself are substantially less than the respective systematic uncertainties listed in the table. All values are expressed relative to full-scale range, or, in the case of ADC input noise, the measured rms value.

Overall System Performance

For purposes of performance verification and quality control, test units which have been independently developed and characterized are particularly well suited, provided they are of sufficient accuracy and stability. While commercially available data converters cannot yet reliably meet and maintain the <2 ppm limit of linearity error required for such service, a high-speed, low-noise 18-bit DAC developed at NBS is well suited for this purpose [7]. While optimized for low noise

and fast settling, DAC-18 exhibits excellent static behavior as well. While this satisfies the requirements for a DAC check standard, there remained a need to provide similar quality control for ADC testing. In particular, it must be verified that the transition-locking feedback loop in fact produces an input voltage whose average value, as measured by the remaining test set circuitry, is exactly equal to the defined transition level. These assumptions are not tested when calibrating DAC's alone. Again, commercially available ADC's are not adequately well behaved for this task.

To provide a check standard for the ADC operating mode, DAC-18 is combined with a precision analog comparator (developed at NBS for this purpose by H. K. Schoenwetter [7]) to create a precision, simulated ADC. As far as the test set is concerned, the resulting instrument behaves exactly as would an 18-bit test ADC, with errors only slightly greater than those of DAC-18 itself.

Overall system performance is monitored through periodic calibration of these check standards. The results of typical linearity error measurements of the ADC check standard are shown in figure 7. Note that the maximum measured error, which includes contributions from both the test set and the check standard, is less than the estimated systematic uncertainty reported in table 1. These check standard calibration records are kept on file, and the maximum and minimum measured errors of each are plotted against time for a convenient control chart. When the control chart indicates that the peak error is approaching the claimed systematic uncertainty, the check standard will be re-adjusted, using an independent self-calibration routine. A subsequent calibration will be performed to verify that the measured errors have been reduced to less than those claimed. If not, then the test set's reference DAC (DAC-20) will be readjusted using its internal self-calibration circuit.

Specifications for Test Converters

To be compatible with the NBS data converter test set, test units must conform to the following general specifications:

- Nominal resolution from 12 to 18 bits.

- Conversion rate of at least 10 kHz.

- Binary coding format, including binary sign-magnitude, offset binary, 2's complement, 1's complement, and complemented versions of these.

• TTL compatibility.

• Voltage ranges of 0-5 V, $\pm$5 V, 0-10 V, $\pm$10 V.

Preparation of Test Boards

In general, it is the customer's responsibility to mount integrated circuit, hybrid, or modular test converters on suitable test boards, providing all required trimmer circuits, voltage references, input or output amplifiers, recommended power supply decoupling capacitors, and connectors for interfacing to the input/output lines. Fully self-contained converters need only be fitted with the necessary interfacing connectors. In so doing, the customer gains significant performance advantages while at the same time saving the additional fee which would otherwise be charged by NBS for performing this service. High performance converters are often susceptible to small changes in grounding, routing of dynamic signal lines, capacitive loading, etc. Particularly with ADC's, signal dynamics is quite important, even for static testing, since the converter itself always operates at high speeds. When mounted by the customer, the test converter and its support circuitry can be laid out more closely to the way in which it will be used in practice, as well as to the specific recommendations of its manufacturer. The test results should therefore more closely describe the converter's in situ performance. Detailed type and wiring requirements for the interfacing connectors are available on request.

References

1. T. M. Souders and D. R. Flach, "An Automated Test Set for High Resolution Analog-to-Digital and Digital-to-Analog Converters," IEEE Trans. Instrum. Meas., Vol.IM-28, Dec. 1979.

2. T. M. Souders and D. R. Flach, "A 20-Bit Plus Sign, Relay-Switched D/A Converter," NBS Tech. Note 1105, Oct. 1979.

3. J. J. Corcoran, et.al., " A High Resolution Error Plotter for Analog-to-Digital Converters," IEEE Trans. Instru. Meas. Vol. IM-24, Dec. 1975.

4. T. M. Souders and J. A. Lechner, "A Technique for Measuring the Equivalent RMS Input Noise of A/D Converters," IEEE Trans.Instrum. Meas., Vol. IM-29, Dec. 1980.

5. L. F. Pau, "Fast Testing and Trimming of High Accuracy Data Converters," Proceedings AUTOTESTCON '78, IEEE Catalog 78CH 1416-7, Nov. 1978.

6. N. J. Fine, "The Walsh Functions," Encyclopedic Dictionary of Physics, Pergamon Press, Oxford, 1969.

7. H. K. Schoenwetter, "A High Speed Low-Noise 18-Bit Digital-to-Analog Converter," IEEE TRans. Instrum. Meas., Vol. IM-27, Dec. 1978.

8. H. K. Schoenwetter, "A Sensitive Analog Comparator," to be published as NBS Tech. Note.

Table 1

Parameter	Estimated Systematic Uncertainty			
	DAC's		ADC's	
Linearity Error	2.7 ppm	+ 0.04 LSB	4.7 ppm	+ 0.16 LSB
Differential Linearity Error	3.2 ppm	+ 0.04 LSB	5.2 ppm.	+ 0.16 LSB
Offset Error[a]	3 ppm		3 ppm	+ 0.07 LSB
Gain Error[a]	6 ppm		6 ppm	+ 0.13 LSB
RMS Input Noise[b]		-------	-100%; +(20% + 10 μV) Noise introduced by Test Set is approx. 30 nV/$\sqrt{Hz}$ in a 1 MHz BW.	

[a] Measured upon special request only, and only if no adjustable trimmers are provided for these parameters.

[b] Since the effective bandwidth of the test converter is generally unknown, the noise contribution from the test set cannot be determined. Therefore, only an upper limit can be accurately placed on the noise measurements.

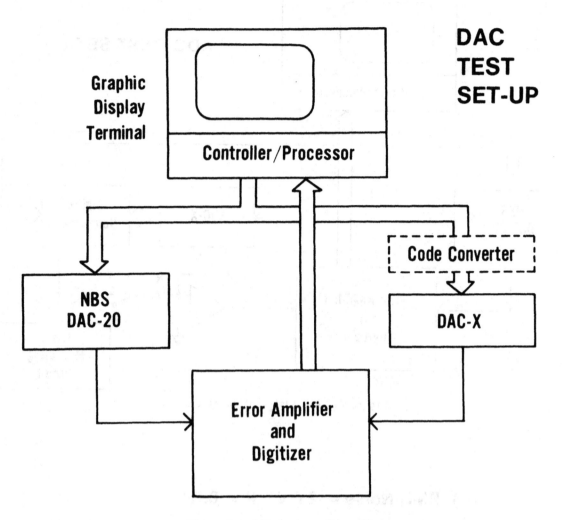

Figure 1. DAC test configuration

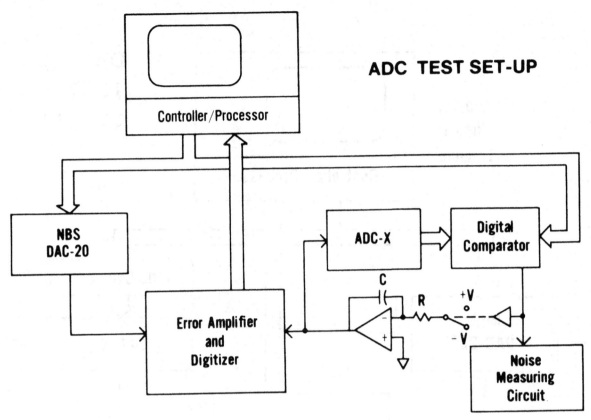

ADC TEST SET-UP

Figure 2. ADC test configuration

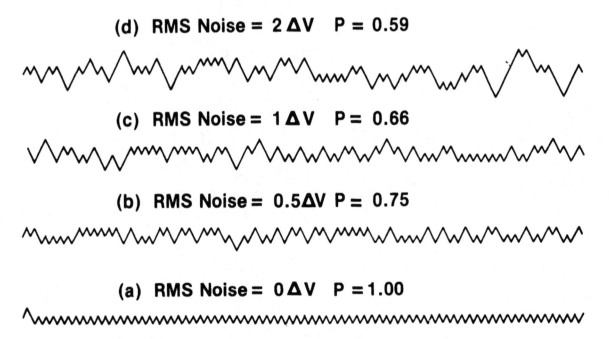

(d) RMS Noise = 2 ΔV   P = 0.59

(c) RMS Noise = 1 ΔV   P = 0.66

(b) RMS Noise = 0.5 ΔV P = 0.75

(a) RMS Noise = 0 ΔV   P = 1.00

Figure 3. Noise response of feedback loop:
converter input versus clock periods

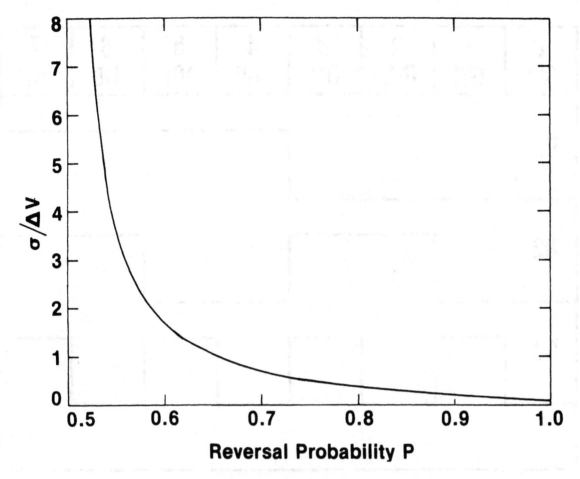

Figure 4. Graph of relationship between the noise
ratio σ/ΔV and the slope reversal
probability P

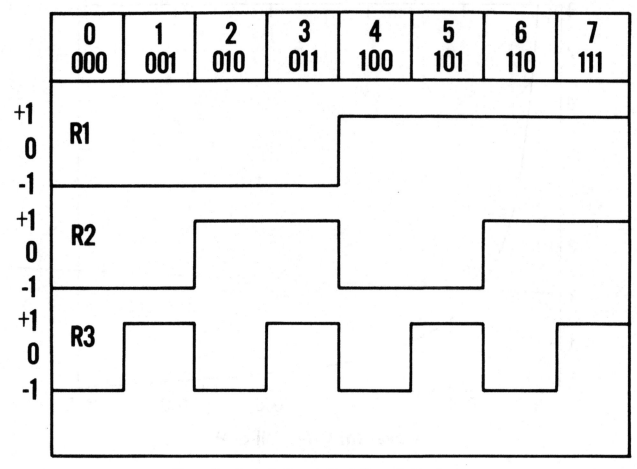

Figure 5. Periodicity of first three Rademacher functions

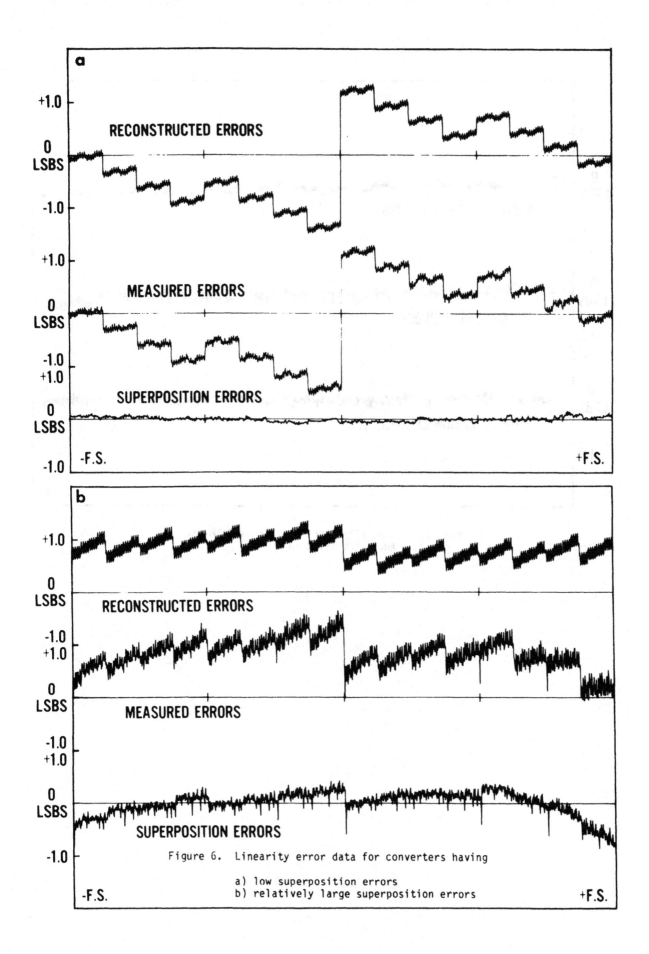

Figure 6. Linearity error data for converters having
  a) low superposition errors
  b) relatively large superposition errors

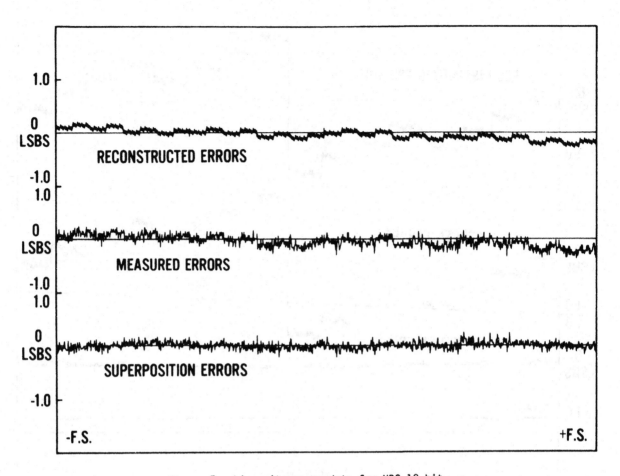

Figure 7. Linearity error data for NBS 18-bit
ADC check standard

PRODUCTION TESTING OF PCM (DIGITAL) AUDIO CIRCUITS

Mark Landry
LTX CORPORATION
LTX Park at University Avenue
Westwood, MA 02090

ABSTRACT

The development of PCM (Digital) audio circuits that are capable of translating digital signals to high quality audio, with noise greater than 96dB down, is challenging the limits of the ATE Industry, both in performing the measurements and in generating the test signals.

This paper examines the various methods available for testing low noise PCM audio circuits. It recommends a new test technique which allows noise measurements below -96dB in high-volume production testing.

PCM AUDIO CIRCUITS

A PCM Audio Circuit (fig. 1) typically includes a 16 Bit D/A converter (DAC), a de-glitcher circuit and a reconstruction filter. The typical test input to the DAC is a digital pattern representing a sinusoid, its output is a stepped sinusoid, either return-to-zero or non-return-to-zero. The de-glitcher circuit will remove the switching glitches of the DAC. The reconstruction filter is one which sharply rolls off at about 20kHz.

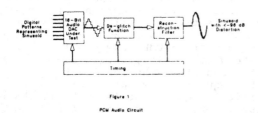

Figure 1

PCM Audio Circuit

There are a number of concerns that must be addressed in testing PCM Audio Circuits. On the measurement side, both distortion and signal to noise must be measured to better than 96dB down, a difficult task for high-volume production test equipment. In signal generation, the test system must be able to generate digital test patterns that produce ideal signals. In both cases, for signal measurement and generation, the production environment creates a conflict between high-speed testing and accurate measurements. With the anticipated high volume production of PCM Audio Circuits and the desired high yields, one must be able to make highly accurate measurements in a high-speed production environment.

THREE METHODS OF TESTING PCM AUDIO DEVICES

1. Precision Notch Filter with an Audio Voltmeter
2. Precision Spectrum Analyzer/Processor
3. Combination of Notch Filter and Digitizer

1. Precision Notch Filter Voltmeter Method

This method uses a precision notch filter combined with an audio voltmeter (fig. 2). This technique is used in many benchtop setups available today.

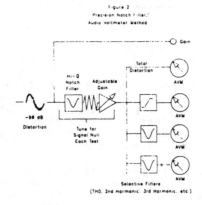

Figure 2
Precision Notch Filter/
Audio Voltmeter Method

In using this method the notch filter must be tuned to totally notch out the fundamental frequency of the sine wave, leaving the noise and distortion products to be measured with an audio voltmeter. A more sophisticated instrument could pass the noise and distortion products through selective filters to read individual components of the distortion product.

Reprinted from *1983 International Test Conference*, 1983, pages 767-770.
Copyright © 1983 by The Institute of Electrical and Electronics Engineers, Inc.

This is a relatively inexpensive way of making extremely high accurate measurements, but it is much too slow for production testing. The precision notch filter, perhaps 100 dB deep, will require many seconds to settle and the filter will have to be continuously tuned to the fundamental frequency, either manually or under microprocessor control. Even under microprocessor control the retuning to maintain stability and accuracy will take more time than can be allowed for production testing.

A modification to this method (fig. 3) is to use a limited depth notch filter in conjunction with a high-pass filter along with the selective filters. The notch filter could be a low-Q filter, about 30-35dB. This filter could settle rapidly and would be stable enough not to require continuous retuning of the notch. The high pass filter would then remove the remainder of the fundamental signal. This method would be faster than the precision notch filter method, but it is still too slow for production testing since each measurement of the selective filters would have to be made separately.

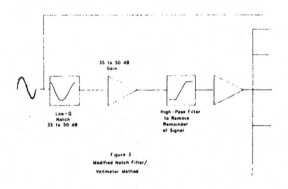

Figure 3
Modified Notch Filter/
Voltmeter Method

2. Precision Spectrum Analyzer Method

The second method for testing PCM Audio Circuits involves using a precision spectrum analyzer/processor (fig. 4) to measure the distortion of the signal. This method is also difficult, perhaps more so than the notch filter method.

Here are two test methods which use a spectrum analyzer. The first involves a digitizer with an internal Fast Fourier Transform processor. It is fast but the distortion measurement is usually accurate to only 85dB at best. A more accurate measurement would require the use of a fast and stable 18 bit A/D converter, which is a very difficult device to produce. The second method is to use a swept-oscillator heterodyne with a bandpass filter. This method could obtain -90dB or so but would again be much too slow for production use.

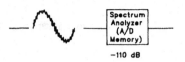

−110 dB

Figure 4

Precision Spectrum Analyzer Method

3. Combination Low-Q Filter Digitizer Method

The third method, and the one recommended by the authors, is a combination of the first two methods. It involves both a notch filter and a digitizer to test PCM audio devices (fig. 5). In this method, however, neither has to be a high-precision element. The digitizer can be a production quality digitizer, capable of analyzing signal to noise in the -75 to -85dB range and the notch filter can be a fast settling, stable, low-Q filter.

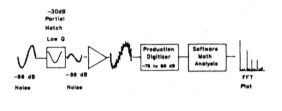

Figure 5

Combination notch filter/digitizer technique

This method lets the notch filter and digitizer share the burden of making the measurements. The filter only needs to be 30 to 35dB down, at the fundamental, instead of 100dB down. The digitizer, instead of 18 bits, only needs to be 12 bits accurate. The low-Q notch filter will not require constant tuning; its only critical requirement is that it have low distortion. Its function is to reduce the size of the fundamental signal without affecting the noise component. Once the signal is digitized, the Fast Fourier Transform (FFT) can be used to obtain measurements of total harmonic distortion, signal to noise ratio, second harmonic and third harmonic, as well as quantization noise introduced by the PCM circuit.

This method is one which can be easily built and maintained in a production environment. The notch filter can be built very close to the device under test and it can be software calibrated, so that the exact shape of the notch is known.

DIGITAL SIGNAL PROCESSING SOFTWARE

Once the output of the PCM DAC is digitized, it is possible to perform matrix math on the digitized data to obtain all the necessary information about the frequency components of the output. The Fast Fourier Transform can be performed on the digitized data with one simple matrix math command:

$$MAT \ Y = FFT(X)$$

where, X is the array containing
the digitized signal
Y contains the sin and cos
frequency components
of the digitized signal

From array Y , specific information about the different frequency components of the signal can be examined. For example, to calculate the TOTAL HARMONIC DISTORTION (THD) of the signal, the power of the fundamental frequency component is compared to the power of the harmonic frequency components. In order to calculate the SIGNAL TO NOISE RATIO (S/N), the power of the fundamental frequency is compared to the power of all the other frequency components of the signal.

The real power of the DSP software is that a great deal of information can be obtained with only one set of digitized data. The matrix math commands can be performed either by the system computer or by an array processor for faster test times.

RESULTS USING THE NOTCH FILTER DIGITIZER TECHNIQUE

If an ideal sine wave is to be tested for distortion, the sine wave can be digitized and DSP software used to obtain the desired results. But if the noise level of the production digitizer is not low enough, the true measurement of distortion cannot be made. This is shown in fig. 6. A 1 kHZ signal, produced by the audio source, is passed through a bandpass filter (to clean the signal to better than -110 dB distortion) and then digitized with a production digitizer. The FFT is then performed and the data is plotted. From the plot (fig. 6) it appears that the 1st and 2nd harmonics are approximately 90 dB down. It is not clear whether this is the true distortion of the signal, or if exceeds the limit of the production digitizer.

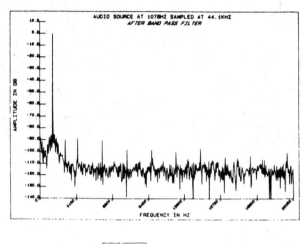

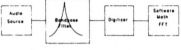

Figure 6

If a calibrated notch filter is added to the measurement circuit and the signal is again digitized, the FFT calculated, and the data plotted (fig. 7), it can be determined from the plot that the harmonics of the signal are better than 110dB down and the noise level of the measurement is about 130dB down.

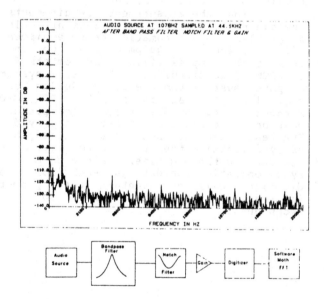

Figure 7

When the calibrated notch filter is added to the production digitizer, as in fig. 7, the noise level of the digitizer is improved and distortion measurements can be made to better than 100 dB down. This proves that the combination of a notch filter and a digitizer creates a test solution capable of testing PCM audio circuits. The plot in fig. 8 is that of the frequency spectrum of a PCM DAC tested with the combination notch filter/digitizer technique. It can be seen from the plot that the noise level is 120dB down and the 1st harmonic is better than 96dB down. Very good for a PCM DAC.

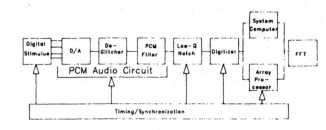

Figure 9

Overall PCM Audio Test Solution

CONCLUSION

It is possible to test PCM audio circuits at production test speeds by using the low-Q notch filter/production digitizer technique. This technique, along with high level DSP software, is capable of making harmonic distortion and signal to noise measurements to better than -100dB. All this is possible at production speed only if the test system allows proper synchronization between the device under test and the test system hardware.

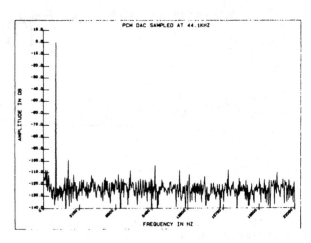

Figure 8

PROVIDING THE DIGITAL STIMULUS

Besides the measurement requirements, the digital stimulus requirements are also a problem for automatic test equipment (fig. 9). The test equipment must be able to meet the complex timing requirements for the PCM audio circuits. The test system computer must be capable of producing the digital input pattern for the PCM DAC without adding any distortion. The pattern must be as accurate as 16 bits can provide. The test equipment should also be capable of synchronizing the test circuitry, the PCM DAC and the digitizer. Without proper synchronization the data gathered from the digitizer will not contain the proper information for using DSP software techniques.

Chapter 14
Appendix: References/
Bibliography

Chapter 14: Appendix: References/Bibliography

I: Reference Articles and Texts for DSP Techniques

[1] B.M. Gordon, "Digital Sampling and Recovery of Analog Signals," *EEE Magazine*, May 1970.

[2] B. Blesser, "Digitization of Audio," *Journal of the A.E.S.*, Oct. 1978, Vol. 26, No. 10, pp. 739-771.

[3] R. Talimbiras, "Some Considerations in the Design of Wide-Dynamic-Range Audio Digitizing Systems," *Proceedings of the A.E.S.*, 57th Convention, May 1977.

[4] M. Mahoney, "The Third Domain of Device Testing," *Electronics Test*, July 1984, Vol. 7, No. 7, pp. 72-84.

[5] F. Brglez, "Digital Signal Processing Considerations in Filter-Codec Testing," *Proceedings of the IEEE 1981 Test Conference*, Computer Society of the IEEE, Washington, D.C., Oct. 1981, pp. 193-200.

[6] R.W. Ramirez, "*The FFT—Fundamentals and Concepts*," Prentice-Hall, Englewood Cliffs, N.J., 1985.

[7] S. Stearns, *Digital Signal Analysis*, Hayden Book Co., N.J., 1975.

[8] R. Rabiner and C. Rader, *Digital Signal Processing*, IEEE Press, New York, 1972.

[9] A. Oppenheim and R. Schafer, *Digital Signal Processing*, Prentice-Hall, Inc., Englewood Cliffs, N.J., 1975.

[10] L. Rabiner and B. Gold, *Theory and Application of Digital Signal Processing*, Prentice-Hall Inc., Englewood Cliffs, N.J., 1975.

[11] R. Crochiere and L. Rabiner, *Multirate Digital Signal Processing*, Prentice-Hall, Inc., Englewood Cliffs, N.J., 1983.

[12] "Transmission Parameters . . . ," *Bell System Publication 41009*, AT&T, Basking Ridge, N.J., May 1975.

[13] *Red Book*, Volume III-3, International Telecommunications Union, Geneva, 1984.

[14] R.S. Burington and D.C. May, *Handbook of Probability and Statistics with Tables*," Second Edition, McGraw-Hill, New York, 1970.

References [1] through [5] deal with practical applications of instrumentation and data acquisition. References [6] through [11] develop the mathematical concepts of digital signal processing. For someone new to DSP, Stearns' book [7] provides a clear and uncomplicated introduction to the fundamental mathematical techniques. References [9 to 11] provide greater depth. Reference [8], an IEEE Press publication, is a valuable collection of pioneering papers in the field, and a rich reference source of earlier works. References [12] and [13] are PCM (codec) telephone standards.

II: Reference Articles and Texts for A/D/A Testing

[15] T.M. Souders, "A Dynamic Test Method for High-Resolution A/D Converters," *IEEE Transactions on Instrumentation and Measurement*, March 1982, Vol. IM-31, No. 1, pp. 3-5.

[16] C. Clayton, et al., "FFT Performance Testing of Data Acquisition Systems," *IEEE Transactions on Instrumentation and Measurement*, June 1986, Vol. IM-35, No. 2, pp. 212-221.

[17] W. Gans, "The Measurement and Deconvolution of Time Jitter in Equivalent-Time Waveform Samplers," *IEEE Transactions on Instrumentation and Measurement*," March 1983, Vol. IM-32, No. 1, pp. 126-133.

[18] "Military Specification, Microcircuits, Linear Video A/D Converters," *MIL-M-38510 Rev 5*, Rome Air Development Center, Griffis AFB, N.Y., March 1983.

[19] J. Doernberg, et al., "Full-Speed Testing of A/D Converters," *IEEE Journal of Solid-State Circuits*, Dec. 1984, Vol. SC-19, No. 6, pp. 820-827.

[20] "Dynamic Performance Testing of A/D Converters," *Hewlett-Packard Product Note 5180A-2*.

[21] H. Kitayoshi, et al.,"DSP Synthesized Signal Source for Analog Testing Stimulus and New Test Method," *Proceedings of the IEEE International Test Conference*, Computer Society of the IEEE, Washington, D.C., 1985, pp. 825-834.

[22] K. Uchida, " Testing the Dynamic Performance of High-Speed A/D Converters," *Proceedings of the IEEE International Test Conference*, Computer Society of the IEEE, Washington, D.C., 1982, pp. 435-440.

[23] E. Sloane, "A System for Converter Testing Using Walsh Transform Techniques," *Proceedings of the IEEE 1981 Test Conference*, Computer Society of the IEEE, Washington, D.C., 1981, pp 304–311.

[24] M. Mahoney, "Envelope and Group Delay," *LTX Technical Topics*, Westwood, Mass., No. 4, 1985

[25] T. Souders, "Modeling and Test Point Selection for Data Converter Testing," *Proceedings of the IEEE International Test Conference*, Computer Society of the IEEE, Washington, D.C., 1985, pp. 813–817.

[26] D. Sheingold and the Analog Devices' Engineering Staff, *Analog Digital Conversion Handbook*, Third Edition, Prentice-Hall, Inc., Englewood Cliffs, N.J., 1986.

[27] B. Gordon, *The Analogic Data-Conversion Systems Digest*, Fourth Edition, Analogic Corp., Wakefield, Mass., 1981.

Bibliography

M. R. Aaron and J. F. Kaiser, "On the calculation of transient response," *Proc. IEEE*, vol. 53, Sept. 1965, pp. 1269.

M. R. Aaron, R. A. McDonald, and E. N. Protonotarios, "Entropy power loss in linear sampled data filters," *Proc. IEEE*, vol. 55, June 1967, pp. 1093-1094.

T. A. Abele, "Transmission factors with Chebyshev-type approximations of constant group delay," *Arch. Elek. Ubertragung*, vol. 16, pp. 9-18, Jan. 1962.

V. R. Algazi, "Useful approximations to optimum quantization," *IEEE Trans. Commun. Technol.*, vol. COM-14, pp. 297-301, June 1966.

A. S. Alivi and G. M. Jenkins, "An example of digital filtering," *Appl. Statistics*, M116-AD, pp. 70-74.

L. E. Alsop and A. A. Nowroozi, "Fast Fourier analysis," *J. Geophys. Res.*, vol. 71, pp. 5482-5483, Nov. 15, 1966.

S. B. Alterman, "Discrete-time least squares, minimum mean square error, and minimax estimation," *IEEE Trans. Commun. Technol.*, vol. COM-14, pp. 302-308, June 1968.

E. B. Anders, "An error bound for a numerical filtering technique," *J. Ass. Comput. Mach.*, vol. 12, pp. 136-140, Jan. 1965.

E. B. Anders et. al., "Digital filters," NASA Contractor Rep. CR-136, Dec. 1964.

H. Andrews, "A high-speed algorithm for the computer generation of Fourier transforms," *IEEE Trans. Comput.* (Short note), vol. C-17, pp. 373-375, Apr. 1968.

C. B. Archambeau et. al., "Data processing techniques for the detection and interpretation of teleseismic signals," *Proc. IEEE*, vol. 53, Dec. 1965, pp. 1860-1884.

C. R. Arnold, "Laguerre functions and the Laguerre Network—their properties and digital simulation," Lincoln Lab., Lexington, Mass., Rep, 1966-28, May 4, 1966.

J. A. Athanassopoulos and A. D. Warren, "Design of discrete-time systems by mathematical programming," in *Proc. 1968 Hawaii Int. Conf. System Sci.* Honolulu: Univ. Hawaii Press, 1968, pp. 224-227.

——, "Time-domain synthesis by nonlinear programming," in *Proc. 4th Allerton Conf.*, 1968, pp. 766-775.

M. L. Attansoro D'atri, and T. Cianciolo, "Pole sensitivity to coefficient rounding in digital filters with multiple shift sequences," *Electron. Lett.*, vol. 7, pp. 29-31, Jan. 1971.

E. Avenhaus and W. Schüssler, "On the approximation problem in the design of digital filters with limited wordlength," *Arch. Elek. Übertragung*, vol. 24, pp. 571-572, 1970.

J. S. Bailey, "A fast Fourier transform without multiplications," in *Proc. Polytechnic Inst. Brooklyn Symp. Comput. Process. Commun.*, 1969.

D. C. Baxter, "The digital simulation of transfer functions," Nat. Res. Labs., Ottawa, Canada, DME Rep. MK-13, Apr. 1964.

——, "Digital simulation using approximate methods," Nat. Res. Council, Ottawa, Canada, Rep. MK-15, July 1965.

G. A. Bekey, "Sensitivity of discrete systems to variation of sampling interval," *IEEE Trans. Automat. Contr.*, vol. AC-11, pp. 284-287, Apr. 1966.

V. A. Benignus, "Estimation of the coherence spectrum and its confidence interval using the fast Fourier transform," *IEEE Trans. Audio Electroacoust.*, vol. AU-17, pp. 145-150, June 1969.

A. Bennett and A. Sage, "Discrete system sensitivity and variable increment sampling," *Proc. Joint Automat. Contr. Conf.*, pp. 603-612, 1967.

W. R. Bennett, "Spectra of quantized signals," *Bell Syst. Tech. J.*, vol. 27, pp. 446-472, July 1948.

G. D. Bergland and R. Klahn, "Digital processor for calculating Fourier coefficients," U.S. Patent 3 544 775, Dec. 29, 1966.

G. D. Bergland, "The fast Fourier transform recursive equations for arbitrary length records," *Math. Comput.*, vol. 21, pp. 236-238, 1967.

——, "A fast Fourier transform algorithm using base 8 iterations," *Math. Comput.*, vol. 22, pp. 275-279, Apr. 1968.

——, "A guided tour of the fast Fourier transform," *IEEE Spectrum*, vol. 6, pp. 41-52, July 1969.

——, "Fast Fourier transform hardware implementations—a survey," *IEEE Trans. Audio Electroacoust.*, vol. AU-17, pp. 109-119, June 1969.

——, "A radix-eight fast Fourier transform subroutine for real-valued series," *IEEE Trans. Audio Electroacoust.*, vol. AU-17, pp. 138-144, June 1969.

——, "Fast Fourier transform hardware implementations—an overview," *IEEE Trans. Audio Electroacoust.*, vol. AU-17, pp. 104-108, June 1969.

G. D. Bergland and H. W. Hale, "Digital real-time spectral analysis," *IEEE Trans. Electron. Comput.*, vol. EC-16, pp. 180-185, Apr. 1967.

G. D. Bergland and D. E. Wilson, "An FFT algorithm for a global, highly parallel processor," *IEEE Trans. Audio Electroacoust.*, vol. AU-17, pp. 125-127, June 1969.

J. E. Bertram, "The effect of quantization in sampled-feedback systems," *AIEE Trans.*, vol. 77, p. 177, 1958.

S. Bertram, "Frequency analysis using the discrete Fourier transform," *IEEE Trans. Audio Electroacoust.*, vol. AU-18, pp. 495-500, Dec. 1970.

——, "On the derivation of the fast Fourier transform," *IEEE Trans. Audio Electroacoust.*, vol. AU-18, pp. 55-58, Mar. 1970.

Bibliography, *Geophysics*, vol. 32, pp. 522-525, June 1967.

C. Bingham, M. D. Godfrey, and J. W. Tukey, "Modern techniques of power spectrum estimation," *IEEE Trans. Audio Electroacoust.*, vol. AU-15, pp. 56-66, June 1967.

L. I. Bluestein, "A linear filtering approach to the computation of discrete Fourier transform," *IEEE Trans. Audio Electroacoust.*, vol. AU-18, pp. 451-455, Dec. 1970.

M. Blum, "On exponential digital filters," *J. Ass. Comput. Mach.*, vol. 6, pp. 283-304, Apr. 1959.

B. Bogert, M. Healy, and J. Tukey, "The quefrency alanysis of time series for echoes," in *Proc. Symp. Time Series Analysis*, M. Rosenblatt, Ed. New York: Wiley, 1963, pp. 209-243.

B. Bogert and E. Parzen, "Informal comments on the uses of power spectrum analysis," *IEEE Trans. Audio Electroacoust.*, vol. AU-15, pp. 74-76, June 1967.

R. E. Bogner, "Frequency sampling filters—Hilbert transformers and resonators," *Bell Syst. Tech. J.*, vol. 48, pp. 501-510, Mar. 1969.

J. Boothroyd, "Complex Fourier series," *Comput. J.*, vol. 10, pp. 414-416, Feb. 1968.

E. M. Boughton, "Definition and synthesis of optimum-smoothing processes in filter terms," *IRE Trans. Instrum.*, vol. I-7, pp. 82-90, Mar. 1958.

R. Boxer, "Frequency analysis of computer systems," *Proc. IRE* (Corresp.), vol. 43, Feb. 1955, pp. 228-229.

——, "A note on numerical transform calculus," *Proc. IRE*, vol. 45, Oct. 1957, pp. 1401-1406.

R. Boxer and S. Thaler, "A simplified method of solving linear and nonlinear systems," *Proc. IRE*, vol. 44, Jan. 1956, pp. 89-101.

N. M. Brenner, "Fast Fourier transform of externally stored data," *IEEE Trans. Audio Electroacoust.*, vol. AU-17, pp. 128-132, June 1969.

——, "Three Fortran programs that perform the Cooley-Tukey Fourier transform," Lincoln Lab., Massachusetts Inst. Technol., Lexington, Tech. Note 1967-2, July 28, 1967.

E. O. Brigham and R. E. Morrow, "The fast Fourier transform," *IEEE Spectrum*, vol. 4, pp. 63-70, Dec. 1967.

H. W. Briscoe and P. L. Fleck, "A real-time computing system for LASA," in *1966 Spring Joint Computer Conf., AFIPS Conf. Proc.*, vol. 28. Washington, D.C.: Spartan, 1966.

J. Brogan, "Filters for sampled signals," *Proc. Symp. Networks Polytechnic Inst. Brooklyn*, Apr. 12-14, 1959, pp. 71-83.

P. W. Broome, "Discrete orthonormal sequences," *J. Ass. Comput. Mach.*, vol. 12, pp. 151-168, Apr. 1965.

P. W. Broome and W. C. Dean, "Seismic applications of orthogonal expansions," *Proc. IEEE*, vol. 53, Dec. 1965, pp. 1865-1869.

P. W. Broome, "A frequency transformation for numerical filters," *Proc. IEEE*, vol. 54, Feb. 1966, pp. 326-327.

J. D. Bruce, "Digital signal processing concepts," *IEEE Trans. Audio Electroacoust.*, vol. AU-18, pp. 344-353, Dec. 1970.

A. Budak and P. Aronhime, "Maximally flat low-pass filters with steeper slopes at cutoff," *IEEE Trans. Audio Electroacoust.*, vol. AU-18, pp. 63-66, Mar. 1970.

H. L. Buijs, "Fast Fourier transformation of large arrays of data," *Appl. Opt.*, vol. 8, pp. 211-212, Jan. 1969.

C. S. Burrus and T. W. Parks, "Time domain design of recursive digital filters," *IEEE Trans. Audio Electroacoust.*, vol. AU-18, pp. 137-141, June 1970.

Reprinted from *Digital Signal Processing*, edited by R. Rabiner and C. Rader, 1972. Copyright © 1972 by The Institute of Electrical and Electronics Engineers, Inc.

A. M. Bush and D. C. Fielder, "An alternative derivation of the z-transform," *Amer. Math. Mon.*, vol. 70, pp. 281–284, Mar. 1963.

V. Cappellini, "Design of some digital filters with application to spectral estimation and data compression," in *Proc. Polytechnic Inst. Brooklyn Symp. Comput. Process. Commun.*, 1969.

——, "Digital filtering with sampled signal spectrum frequency shift," *Proc. IEEE* (Lett.), vol. 57, Feb. 1969, pp. 241–242.

V. Cappellini and T. D'Amico, "Some numerical filters obtained from the evaluation of the convolution integral and their application to spectral analysis," *Alta Freq.*, vol. 36, pp. 835–840, 1967.

C. C. Carroll and R. White, "Discrete compensation of control systems with integrated circuits," *IEEE Trans. Automat. Contr.*, vol. AC-12, pp. 579–582, Oct. 1967.

R. K. Cavin, C. H. Ray, and V. T. Rhyne, "The design of optimal convolutional filters via linear programming," *IEEE Trans. Geosci. Electron.*, vol. GE-7, pp. 142–145, July 1969.

R. K. Cavin and M. C. Budge, Jr., "A note on multirate z-transforms," *Proc. IEEE* (Lett.), vol. 58, Nov. 1970, pp. 1840–1841.

D. Chanoux, "Synthesis of recursive digital filters using the FFT," *IEEE Trans. Audio Electroacoust.*, vol. AU-18, pp. 211–212, June 1970.

E. W. Cheney and H. L. Loeb, "Generalized rational approximation," *SIAM J. Num. Analysis*, vol. 1, pp. 11–25, 1964.

V. Cizek, "Discrete Hilbert transform," *IEEE Trans. Audio Electroacoust.*, vol. AU-18, pp. 340–343, Dec. 1970.

——, "Numerische Hilbert-transformation," *Proc. Inst. Radio Eng. Electron.*, Czechoslovak Academy of Sciences, no. 11, 1961.

J. F. Claerbout and E. A. Robinson, "The error in least-squares inverse filtering," *Geophysics*, vol. 29, pp. 118–120, 1964.

W. T. Cochran et. al., "What is the fast Fourier transform?", *IEEE Trans. Audio Electroacoust.*, vol. AU-15, pp. 45–55, June 1967.

A. G. Constantinides, "Design of bandpass digital filters," *Proc. IEEE*, vol. 57, June 1969, pp. 1229–1231.

——, "Digital filters with equiripple passbands," *IEEE Trans. Circuit Theory* (Corresp.), vol. CT-16, pp. 535–538, Nov. 1969.

——, "Elliptic digital filters," *Electron. Lett.*, vol. 3, pp. 255–256, June 1967.

——, "Frequency transformations for digital filters," *Electron. Lett.*, vol. 3, pp. 487–489, Nov. 1967.

——, "Spectral transformation for digital filters," *Proc. Inst. Elec. Eng.*, vol. 117, 1970, pp. 1585–1590.

——, "Synthesis of Chebychev digital filters," *Electron Lett.*, vol. 3, pp. 124–127, Mar. 1967.

J. W. Cooley, "Complex finite Fourier transform subroutine," Share Doc. 3465, Sept. 8, 1966.

——, "Application of the fast Fourier transform method," in *Proc. IBM Sci. Comput. Symp.*, 1966.

——, "Harmonic analysis of complex Fourier series," Share Program Library No. SDA 3425, Feb. 7, 1966.

J. W. Cooley et. al., "The 1968 Arden house workshop on fast Fourier transform processing," *IEEE Trans. Audio Electroacoust.*, vol. AU-17, pp. 66–76, June 1969.

J. W. Cooley, P. A. Lewis, and P. D. Welch, "Application of the fast Fourier transform to computation of Fourier integrals, Fourier series, and convolution integrals," *IEEE Trans. Audio Electroacoust.*, vol. AU-15, pp. 79–84, June 1967.

——, "The fast Fourier transform and its applications," IBM Res. Paper RC-1743, Feb. 9, 1967.

——, "The fast Fourier transform algorithm: programming considerations in the calculation of sine, cosine, and Laplace transforms," *J. Sound. Vib.*, vol. 12, pp. 315–337, 1970.

——, "The finite Fourier transform," *IEEE Trans. Audio Electroacoust.*, vol. AU-17, pp. 77–86, June 1969.

——, "Historical notes on the fast Fourier transform," *IEEE Trans. Audio Electroacoust.*, vol. AU-15, pp. 76–79, June 1967.

——, "The use of the fast Fourier transform algorithm for the estimation of spectra and cross spectra," in *Proc. Polytechnic Inst. Brooklyn Symp. Comput. Process. Commun.*, 1969.

J. W. Cooley and J. W. Tukey, "An algorithm for the machine computation of complex Fourier series," Share Doc. 3465, Sept. 8, 1966.

M. J. Corinthios, "A fast Fourier transform for high-speed signal processing," *IEEE Trans. Comput.*, vol. C-20, pp. 843–846, Aug. 1971.

——, "A time-series analyzer," in *Proc. Polytechnic Inst. Booklyn Symp. Comput. Process. Commun.*, 1969.

——, "The design of a class of fast Fourier transform computers," *IEEE Trans. Comput.*, vol. C-20, pp. 617–623, June 1971.

T. H. Crystal and L. Ehrman, "The design and applications of digital filters with complex coefficients," *IEEE Trans. Audio Electroacoust.*, vol. AU-16, pp. 315–321, Sept. 1968.

E. E. Curry, "The analysis of round-off and truncation errors in a hybrid control system," *IEEE Trans. Automat. Contr.*, vol. AC-12, pp. 601–604, Oct. 1967.

G. C. Danielson and C. Lanczos, "Some improvements in practical Fourier analysis and their application to x-ray scattering from liquids," *J. Franklin Inst.*, pp. 365–380, pp. 435–452, Apr.–May 1942.

J. A. D'Appolito, "A simple algorithm for discretizing linear stationary continuous time systems," *Proc. IEEE*, vol. 54, Dec. 1966, pp. 2010–2011.

A. C. Davies, "Digital filtering of binary sequences," *Electron. Lett.*, vol. 3, pp. 318–319, July 1967.

L. D. Divieti, C. M. Rossi, R. M. Schmid, and A. E. Vereschkin, "A note on computing quantization errors in digital control systems," *IEEE Trans. Automat. Contr.*, vol. AC-12, pp. 622–623, Oct. 1967.

C. L. Dolph, "A current distribution for broadside arrays which optimizes the relationship between beamwidth and side-lobe level," *Proc. IRE*, vol. 34, June 1946, pp. 335–348.

C. J. Drane, "Directivity and beamwidth approximations for large scanning Dolph-Chebyshev arrays," AFCRL Physical Sci. Res. Paper 117, AFCRL-65-472, June 1965.

G. Dumermuth and H. Fluhler, "Some modern aspects in numerical spectrum analysis of multichannel electroencephalographic data," *Med. Elec. Biol. Eng.*, vol. 5, pp. 319–331, 1967.

S. C. Dutta Roy, "On maximally flat sharp cutoff low-pass filters," *IEEE Trans. Audio Electroacoust.*, vol. AU-19, pp. 58–63, Mar. 1971.

P. M. Ebert, J. E. Mazo, and M. G. Taylor, "Overflow oscillations in digital filters," *Bell Syst. Tech. J.*, vol. 48, pp. 2999–3020, Nov. 1969.

R. Edwards and A. Bradley, "Design of digital filters by computers," *Int. J. Numer. Methods Eng.*, vol. 2, pp. 311–333, 1970.

R. Edwards, J. Bradley, and J. Knowles, "Comparison of noise performances of programming methods in the realization of digital filters," in *Proc. Polytechnic Inst. Brooklyn Symp. Comput. Process. Commun.*, 1969.

G. Epstein, "Recursive fast Fourier transforms," in *1968 Fall Joint Computer Conf. AFIPS Conf. Proc.*, vol. 33. Washington D.C.: Thompson, 1968, pp. 141–143.

M. J. Ferguson and P. E. Mantey, "Automatic frequency control via digital filtering," *IEEE Trans. Audio Electroacoust.*, vol. AU-16, pp. 392–398, Sept. 1968.

A. Fettweis, "A general theorem for signal-flow networks, with applications," *Arch. Elek. Übertragung*, vol. 25, pp. 557–561, Dec. 1971.

——, "Digital filter structures related to classical filter networks," *Arch. Elek. Übertragung*, vol. 25, pp. 79–89, 1971.

A. A. Filippini, "Synthesis of cascaded digital filters to achieve desired transfer characteristics," Tech. Rep. AD-699 529, Sept. 1969.

P. E. Fleischer, "Digital realization of complex transfer functions," *Simulation*, vol. 6, pp. 171–180, Mar. 1966.

R. Fletcher and M. J. Powell, "A rapidly convergent descent method for minimization," *Comput. J.*, vol. 6, pp. 163–168, 1963.

M. L. Forman, "Fast Fourier transform technique and its application to Fourier spectroscopy," *J. Opt. Soc. Am.*, vol. 56, pp. 978–990, July 1966.

M. E. Fowler, "A new numerical method for simulation," *Simulation*, vol. 4, pp. 324–330, May 1965.

P. A. Franaszek, "On sampled-data and time varying systems," Commun. Lab., Dept. Elec. Eng., Princeton Univ., Princeton, N.J., Tech. Rep. 11, Oct. 1965.

L. E. Franks, "Power spectral density of random facsimile signals," *Proc. IEEE* (Corresp.), vol. 52, Apr. 1964, pp. 431–432.

D. Fraser, "Associative parallel processing," in *1967 Spring Joint Computer Conf., AFIPS CONF. Proc.*, vol. 30. Washington, D.C.: Thompson, 1967, pp. 471–475.

R. Galpin, "Variable electronic allpass delay network," *Electron. Lett.*, vol. 4, pp. 137–139, 1968.

A. Gelb and P. Palosky, "Generating discrete colored noise from discrete white noise," *IEEE Trans. Automat. Contr.*, vol. AC-12, pp. 148-149, Jan. 1966.

W. M. Gentleman and G. Sande, "Fast Fourier transforms—for fun and profit," in *1966 Fall Joint Computer Conf., AFIPS Conf. Proc.*, vol. 29. Washington, D.C.: Spartan, 1966, pp. 563-578.

W. M. Gentleman, "Matrix multiplication and fast Fourier transforms," *Bell Syst. Tech. J.*, vol. 47, pp. 1099-1103, July-Aug. 1968.

A. J. Gibbs, "An introduction to digital filters," *Aust. Telecommun. Res.*, vol. 3, pp. 3-14, Nov. 1969.

——, "The design of digital filters," *Aust. Telecommun. Res.*, vol. 4, pp. 29-34, 1970.

G. C. Gillete, "The digiphase synthesizer," *Frequency Technol.*, pp. 25-29, Aug. 1969.

M. Gilmartin, Jr. and R. R. Shively, "Digital processor for performing fast Fourier transforms," U.S. Patent 3 517 173, Dec. 29, 1966.

J. A. Glassman, "A generalization of the fast Fourier transform," *IEEE Trans. Comput.*, vol. C-19, pp. 105-116, Feb. 1970.

T. H. Glisson and A. P. Sage, "On discrete and complex representation of real signals," in *Proc. 12th Midwest Symp. Circuit Theory*, Apr. 1969.

T. H. Glisson, C. I. Black, and A. P. Sage, "The digital computation of discrete spectra using the fast Fourier transform," *IEEE Trans. Audio Electroacoust.*, vol. AU-18, pp. 271-287, Sept. 1970.

G. Goertzel, "An algorithm for the evaluation of finite trigonometric series," *Amer. Math. Mon.*, vol. 65, pp. 34-35, Jan. 1958.

B. Gold and K. Jordan, "A direct search procedure for designing finite duration impulse response filters," *IEEE Trans. Audio Electroacoust.*, vol. AU-17, pp. 33-36, Mar. 1969.

——, "A note on digital filter synthesis," *Proc. IEEE*, vol. 56, Oct. 1968, pp. 1717-1718.

B. Gold and L. R. Rabiner, "Analysis of digital and analog formant synthesizers," *IEEE Trans. Audio Electroacoust.*, vol. AU-16, pp. 81-94, Mar. 1968.

B. Gold and C. M. Rader, "Effects of quantization noise in digital filters," in *1966 Spring Joint Computer Conf., AFIPS Conf. Proc.*, vol. 28. Washington, D.C.: Spartan, 1966, pp. 213-219.

B. Gold, A. V. Oppenheim, and C. M. Rader, "Theory and implementation of the discrete Hilbert transform," in *Proc. Polytechnic Inst. Booklyn Symp. Comput. Process. Commun.*, 1969.

R. M. Golden, "Digital computer simulation of sampled-data communication systems using block diagram compiler: BLODIB," *Bell Syst. Tech. J.*, vol. 45, pp. 344-358, Mar. 1966.

——, "Digital computer simulation of sampled-data voice-excited vocoder," *J. Acoust. Soc. Am.*, vol. 35, pp. 1358-1366, 1963.

——, "Digital filter synthesis by sampled-data transformation," *IEEE Trans. Audio Electroacoust.*, vol. AU-16, pp. 321-329, Sept. 1968.

R. M. Golden and J. F. Kaiser, "Design of wideband sampled-data filters," *Bell Syst. Tech. J.*, vol. 43, pp. 1533-1546, July 1964.

——, "Root and delay parameters for normalized Bessel and Butterworth low-pass transfer functions," *IEEE Trans. Audio Electroacoust.*, vol. AU-19, pp. 64-71, Mar. 1971.

R. M. Golden and S. A. White, "A holding technique to reduce the number of bits in digital transfer functions," *IEEE Trans. Audio Electroacoust.*, vol. AU-16, pp. 433-437, Sept. 1968.

I. J. Good, "The interaction algorithm and practical Fourier series," *J. Roy Statist. Soc., Ser. B.*, vol. 20, pp. 361-372, 1958; Addendum, vol. 22, pp. 372-375, 1960.

——, "The relationship between two fast Fourier transforms," *IEEE Trans. Comput.*, vol. C-20, pp. 310-317, Mar. 1971.

D. J. Goodman, "Optimum digital filters for the estimation of continuous signals in noise," *Proc. Polytechnic Inst. Brooklyn Symp. Comput. Process. Commun.*, 1969.

L. M. Goodman and P. R. Drouilhet, Jr., "Asymptotically optimum pre-emphasis and de-emphasis networks for sampling and quantizing," *Proc. IEEE*, vol. 54, May 1966, pp. 795-796.

O. D. Grace, "Two finite Fourier transforms for bandpass signals," *IEEE Trans. Audio Electroacoust.* (Corresp.), vol. AU-18, pp. 501-502, Dec. 1970.

O. D. Grace and S. P. Pitt, "Quadrature sampling of high-frequency waveforms," *J. Acoust. Soc. Am.*, vol. 44, pp. 1453-1454, Nov. 1968.

R. J. Graham, "Determination and analysis of numerical smoothing elements," NASA Tech. Rep. TR-R-179, Dec. 1963.

A. Grassi and G. Strini, "Errors in the reconstruction of quantized and sampled random signals," *Alta Freq.*, vol. 33, pp. 547-555, Aug. 1964.

C. J. Greaves and J. A. Cadzow, "The optimal discrete filter corresponding to a given analog filter," *IEEE Trans. Automat. Contr.*, vol. AC-12, pp. 304-307, June 1967.

B. F. Green, J. E. Smith, and L. Klem, "Empirical tests of an additive random number generator," *J. Ass. Comput. Mach.*, vol. 6, pp. 527-537, Oct. 1959.

H. L. Groginsky and G. A. Works, "A pipeline fast Fourier transform," *IEEE Trans. Comput.*, vol. C-19, pp. 1015-1019, Nov. 1970.

S. C. Gupta and W. W. Happ, "Flowgraph approach to z-transforms and application to discrete systems," *Int. J. Contr.*, vol. 2, pp. 211-220, Sept. 1965.

C. A. Halijak, "Digital approximation of differential equations by trapezoidal convolution," Kansas State Univ. Bulletin, vol. 45, Rep. 10, pp. 83-94, July 1961.

S. R. Harrison and B. J. Leon, "Digital filters," Tech. Rep. N70-20579, Sept. 1969.

W. T. Hartwell and R. A. Smith, "Apparatus for performing complex wave analysis," U.S. Patent 3 544 894, July 10, 1967.

H. S. Heaps and W. Willcock, "The use of quantizing techniques in real time Fourier analysis," *Radio Electron. Eng.*, vol. 29, pp. 143-148, Mar. 1965.

D. Helman, "Tchebycheff approximations for amplitude and delay with rational functions," in *Proc. Polytechnic Inst. Brooklyn Symp. Modern Network Theory*, pp. 385-402.

H. D. Helms, "Digital filters with equiripple or minimax responses," *IEEE Trans. Audio Electroacoust.*, vol. AU-19, pp. 87-94, Mar. 1971.

——, "Fast Fourier transform method of computing difference equations and simulating filters," *IEEE Trans. Audio Electroacoust.*, vol. AU-15, pp. 85-90, June 1967.

——, "Nonrecursive digital filters: design methods for achieving specifications on frequency response," *IEEE Trans. Audio Electroacoust.*, vol. AU-16, pp. 336-342, Sept. 1968.

R. M. Hendrickson, "Frequency response functions of certain numerical filters," Space Technol. Labs., Redondo Beach, Calif., GM-00-4330-00281, Apr. 6, 1959.

O. Herrmann, "Design of nonrecursive digital filters with linear phase," *Electron. Lett.*, vol. 6, pp. 328-329, 1970.

——, "On the approximation problem in nonrecursive digital filter design," *IEEE Trans. Circuit Theory*, vol. CT-18, pp. 411-413, 1971.

O. Herrmann and W. Schüssler, "Design of nonrecursive digital filters with minimum phase," *Electron. Lett.*, vol. 6, 1970.

——, "On the accuracy problem in the design of nonrecursive digital filters," *Arch. Elek. Übertragung*, vol. 24, pp. 525-526, 1970.

G. E. Heyliger, "The scanning function approach to the design of numerical filters," Martin Comp., Denver, Colo., Rep. R-63-2, Apr. 1963.

——, "Simple design parameters for Chebyshev arrays and filters," *IEEE Trans. Audio Electroacoust.* (Corresp.), vol. AU-18, pp. 502-503, Dec. 1970.

F. B. Hills, "A study of incremental computation by difference equation," Servomechanisms Lab., Massachusetts Inst. Technol., Cambridge, Mass., Rep. 7849-R-1, May 1958.

M. J. Hinich and C. S. Clay, "The application of the discrete Fourier transform in estimation of power spectra, coherence and bispectra of geophysical data," *Rev. Geophys.*, vol. 6, pp. 347-363, Aug. 1968.

E. Hofstetter, A. V. Oppenheim, and J. Siegel, "A new technique for the design of nonrecursive digital filters," in *Proc. 5th Annu. Princeton Conf. Inform. Sci. Syst.*, 1971, pp. 64-72.

——, "On optimum nonrecursive digital filters," in *Proc. 9th Annu. Allerton Conf. Circuit System Theory*, Oct. 6-8, 1971.

H. Holtz and C. T. Leondes, "The synthesis of recursive filters," *J. Ass. Comput. Mach.*, vol. 13, pp. 262-280, Apr. 1966.

T. C. Hsia, "On synthesis of optimal digital filters," in *Proc. 1st Asilomar Conf. Circuit System Theory*, Nov. 1-3, 1967, pp. 473-480.

W. H. Huggins, "Signal Theory," *IRE Trans. Circuit Theory*, vol. CT-3, pp. 210-216, Dec. 1956.

L. B. Jackson, "An analysis of limit cycles due to multiplication rounding in recursive digital filters," in *Proc. 7th Annu. Allerton Conf. Circuit System Theory*, 1969, pp. 69-78.

——, "On the interaction of roundoff noise and dynamic range in digital filters," *Bell Syst. Tech. J.*, vol. 49, pp. 159-184, 1970.

——, "Roundoff-noise analysis for fixed-point digital filters realized in cascade or parallel form," *IEEE Trans. Audio Electroacoust.*, vol. AU-18, pp. 107-122, June 1970.

L. B. Jackson, J. F. Kaiser, and H. S. McDonald, "An approach to the implementation of digital filters," *IEEE Trans. Audio Electroacoust.*, vol. AU-16, pp. 413-421, Sept. 1968.

L. B. Jackson and H. S. McDonald, "Digital filtering," U.S. Patent 3 522 546, Aug. 4, 1970.

G. W. Johnson, D. P. Libdorff, and C. G. Nordling, "Extension of continuous-data system design techniques to sampled-data control systems," *AIEE Trans. Appl. Ind.*, vol. 74, part 2, pp. 252-263, Sept. 1955.

N. B. Jones, "Lowpass filters with approximately equal ripple modulus error," *Electron. Lett.*, vol. 3, pp. 516-517, Nov. 1967.

E. I. Jury, "A general z-transform formula for sampled-data systems," *IEEE Trans. Automat. Contr.*, vol. AC-12, pp. 606-608, Oct. 1967.

D. K. Kahaner, "Matrix description of the fast Fourier transform," *IEEE Trans. Audio Electroacoust.*, vol. AU-18, pp. 442–450, Dec. 1970.

R. E. Kahn and B. Liu, "Sampling with time jitter," Commun. Lab., Dept. Elec. Eng., Princeton Univ., Princeton, N.J., Tech. Rep. 8, June 1964.

J. F. Kaiser, "Computer aided design of classical continuous system transfer functions," in *Proc. Hawaii Int. Conf. System Sci.* Honolulu: Univ. Hawaii Press, pp. 197-200, 1968.

___, "Design methods for sampled-data filters," in *Proc. 1st Annu. Allerton Conf. Circuit System Theory*, 1963, pp. 221-236.

___, "Digital filters," in *System Analysis by Digital Computer*, F. F. Kuo and J. F. Kaiser, Ed. New York: Wiley, 1966.

___, "Some practical considerations in the realization of linear digital filters," in *Proc. 3rd Annu. Allerton Conf. Circuit System Theory*, 1963, pp. 621–633.

S. C. Kak, "The discrete Hilbert transform," *Proc. IEEE* (Lett.), vol. 58, Apr. 1970, pp. 585-586.

H. E. Kallmann, "Transversal filters," *Proc. IRE*, vol. 28, July 1940, pp. 302-310.

R. E. Kalman and J. E. Bertram, "A unified approach to the theory of sampled systems," *J. Franklin Inst.*, vol. 267, pp. 405–436, May 1959.

T. Kaneko and B. Liu, "Roundoff error of floating-point digital filters," in *Proc. 6th Annu. Allerton Conf. Circuit System Theory*, 1968, pp. 219-227.

___, "Accumulation of round-off errors in fast Fourier transform," *J. Ass. Comput. Mach.*, vol. 17, pp. 637-654, Oct. 1970.

B. J. Karafin, "The new block diagram compiler for simulations of sampled-data systems," in *1965 Fall Joint Computer Conf., AFIPS Conf. Proc.*, vol. 27. Washington, D.C.: Spartan, 1965, pp. 55–61.

J. Katzenelson, "On errors introduced by combined sampling and quantization," *IRE Trans. Automat. Contr.*, vol. 7, pp. 58–68, Apr. 1962.

___, "A note on errors introduced by combined sampling and quantization," Electron. Systems Lab., Massachusetts Inst. Technol., Cambridge, Mass., ESL-TM-101, Mar. 1961.

W. H. Kautz, "Transient synthesis in the time domain," *IRE Trans. Circuit Theory*, vol. CT-1, pp. 29-39, Sept. 1954.

J. E. Kelley, Jr., "An application of linear programming to curve fitting," *J. Soc. Ind. Appl. Math.*, vol. 6, pp. 15-22, 1968.

W. C. Kellogg, "Information rates in sampling and quantizing," *IEEE Trans. Inform. Theory*, vol. IT-13, pp. 506-511, July 1967.

J. L. Kelly, C. L. Lochbaum, V. A. Vyssotsky, "A block diagram compiler," *Bell Syst. Tech. J.*, vol. 40, pp. 669-676, 1961.

L. C. Kelly and J. N. Holmes, "Computer processing of signals with particular reference to simulation of electric filter networks," Eng. GPO Dept., Post Office Res. Station, London, Eng., Res. Rep. 21072, Feb. 15, 1965.

R. Klahn and R. R. Shively, "FFT—shortcut to Fourier analysis," *Electronics* vol. 41, pp. 124-129, Apr. 15, 1968.

R. Klahn, R. R. Shively, E. Gomez, and M. J. Gilmartin, "The time-saver-FFT hardware," *Electronics*, vol. 41, pp. 92-97, June 24, 1968.

J. B. Knowles and R. Edwards, "Aspects of subrate digital control systems," *Proc. Inst. Elec. Eng.*, vol. 113, Nov. 1966, pp. 1885-1892.

___, "Complex cascade programming and associated computational errors," *Electron. Lett.*, vol. 1, pp. 160-161, Aug. 1965.

___, "Computational error effects in a direct digital control system," *Automatica*, vol. 4, pp. 7-29, 1966.

___, "Effects of a finite word length computer in a sampled-data feedback system," *Proc. Inst. Elec. Eng.*, vol. 112, June 1965.

___, "Finite word-length effects in a multirate direct digital control system," *Proc. Inst. Elec. Eng.*, vol. 112, Dec. 1965, pp. 2376-2384.

J. B. Knowles and E. M. Olcayto, "Coefficient accuracy and digital filter response," *IEEE Trans. Circuit Theory*, vol. CT-15, pp. 31–41 Mar. 1968.

A. Kohlenberg, "Exact interpolation of band-limited functions," *J. Appl. Phys.*, vol. 24, pp. 1432-1436, Dec. 1953.

G. A. Korn, "Hybrid-computer techniques for measuring statistics from quantized data," *Simulation*, vol. 4, pp. 229-239, Apr. 1965.

E. Korngold, "The periodic analysis of sampled data," Lincoln Lab. Massachusetts Inst. Technol., Lexington, Mass., Group Rep. 1964-32 June 15, 1964.

A. A. Kosyakin, "The statistical theory of amplitude quantization," *Avtomat. Telemekh.*, vol. 22, p. 722, 1961.

W. Kuntz, "A new sample-and-hold device and its application to the realization of digital filters," *Proc. IEEE*, vol. 56, Nov. 1968 pp. 2092-2093.

W. Kuntz and H. W. Schüssler, "The numerical calculation of the time response networks with the aid of the z-transformation," *J. Nachricht tentech Z.*, vol. 5, pp. 121-124, 1967.

G. N. Lack, "Comments on upper bound on dynamic quantization error in digital control systems via the direct method of Liapunov," *IEEE Trans. Automat. Contr.* (Corresp.), vol. AC-11, pp. 331-333 Apr. 1966.

I. M. Langenthal, "Coefficient sensitivity and generalized digital filter synthesis," *1968 Eascon Rec.*, pp. 386-392, 1968.

___, "The synthesis of symmetrical bandpass digital filters," in *Proc. Polytechnic Inst. Brooklyn Symp. Comput. Process. Commun.*, 1969

I. M. Langenthal and S. Gowrinathan, "Advanced digital processing techniques," Tech. Rep. AD 708736, 1970.

A. W. Langill, "Digital filters," *Frequency*, vol. 6, Nov.-Dec. 1964.

A. G. Larson and R. C. Singleton, "Real-time spectral analysis on a small general-purpose computer," *1967 Fall Joint Computer Conf. AFIPS Conf. Proc.*, vol. 31. Washington, D.C.: Thompson, 1967 pp. 665-674.

R. M. Lerner, "Band-pass filters with linear phase," *Proc. IEEE*, vol. 52 Mar. 1964, pp. 249-268.

L. B. Lesem, P. M. Hirsch, J. A. Jordan, Jr., "Computer synthesis of holograms for 3-D display," *Commun. Ass. Comput. Mach.*, vol. 11 pp. 661-674, Oct. 1968.

M. J. Levin, "Estimation of a system pulse transfer function in the presence of noise," *IEEE Trans. Automat. Contr.*, vol. AC-9, pp. 229 235, July 1964.

___, "Generation of a sampled Gaussian time series having a specified correlation function," *IRE Trans. Inform. Theory*, vol. IT-6, pp. 545 548, Dec. 1960.

P. M. Lewis, "Synthesis of sampled signal networks," *IRE Trans. Circuit Theory*, vol. CT-5, pp. 74-77, Mar. 1958.

R. N. Linebarger, "Precision-sample rate tradeoffs in quantized sampled data systems," Systems Res. Center, Case Inst. Technol., Cleveland Ohio, Rep. SRC-46-C-64-17, 1964.

W. K. Linvill, "Sampled-data control systems studied through comparison of sampling with amplitude modulation," *AIEE Trans.*, vol. 70, part 2, pp. 1779-1788, 1951.

B. Liu, "Effect of finite word length on the accuracy of digital filters—a review," *IEEE Trans. Circuit Theory*, vol. CT-18, pp. 670-677 Nov. 1971.

B. Liu and P. Franaszek, "A class of time-varying digital filters," *IEEE Trans. Circuit Theory*, vol. CT-16, pp. 467–471, Nov. 1969.

B. Liu and J. B. Thomas, "Error problems in sampling representations part I, type I errors," Commun. Lab., Dept. Elec. Eng., Princeton Univ., Princeton, N.J., Tech. Rep. 4, Z08250, Apr. 1964.

___, "Error problems in the reconstruction of signals from sampled data," *Proc. Nec.*, vol. 23, pp. 803-807, 1967.

P. A. Lynn, "Economic linear-phase recursive digital filters," *Electron. Lett.*, vol. 6, pp. 143-145, Mar. 1970.

M. D. MacLaren and G. Marsaglia, "Uniform random number generators," *J. Ass. Comput. Mach.*, vol. 12, pp. 83-89, 1965.

C. E. Maley, "The effect of parameters on the roots of an equation system, *Comput. J.*, vol. 4, pp. 62-63, 1961-1962.

C. G. Maling, Jr., W. T. Morrey, and W. W. Lang, "Digital determination of third-octave and full-octave spectra of acoustical noise," *IEEE Trans. Audio Electroacoust.*, vol. AU-15, pp. 98-104, June 1967.

R. Manasse, "Tapped delay line realizations of frequency periodic filters and their application to linear FM pulse compression," Mitre Corp., Bedford, Mass., Tech. Doc. Rep. ESD-TDR-63-232, May 1963.

M. Mansour, "Instability criteria of linear discrete systems," *Aut.*

matica, vol. 2, pp. 167–178, Jan. 1965.

. E. Mantey, "Convergent automatic-synthesis procedures for sampled data networks with feedback," Electron. Labs., Stanford Univ., Stanford, Calif., Tech. Rep. 6773-1, SU-SEL-64-112, Oct. 1964.

. E. Mantey and G. F. Franklin, "Digital filter design techniques in the frequency domain," *Proc. IEEE*, vol. 55, Dec. 1967, pp. 2196–2197.

. E. Mantey, "Eigenvalue sensitivity and state-variable selection," *IEEE Trans. Automat. Contr.*, vol. AC-13, pp. 263–269, June 1968.

. Marsaglia, "A note on the construction of a multivarate normal sample," *IRE Trans. Inform. Theory*, vol. IT-3, p. 149, June 1957.

M. A. Martin, "Frequency domain applications to data processing," *IRE Trans. Space Electron. Telem.*, vol. SET-5, pp. 33–41, Mar. 1959

___, "Digital filters for data processing," General Electric Comp., Missile Space Div., Tech. Inform., Series Rep. 62-SD484, 1962.

. Max, "A new Fourier technique for frequency-domain synthesis of delay-line filters," in *Proc. 7th Midwest Symp. Circuit Theory*, May 4–5, 1964, pp. 165–170.

D. W. McCowan, "Finite Fourier transform theory and its application to the computation of convolutions, correlations, and spectra," Teledyne Industries, Inc., Earth Sciences Div., Oct., 1966.

R. N. McDonough, "Comment on 'z-transform' technique," *Proc. IEEE*, vol. 54, Nov. 1966, pp. 1616–1617.

R. N. McDonough and W. H. Huggins, "Best least-squares representation of signals by exponentials," *IEEE Trans. Automat. Contr.*, vol. AC-13, pp. 408–412, Aug. 1968.

. T. McKeever, "The associative memory structure," in *1965 Fall Joint Computer Conf.*, AFIPS Conf. Proc., vol. 27. Washington, D.C.: Spartan, 1965, pp. 371–388.

. H. McKinney, "A digital spectrum channel analyzer," in *Conf. Speech Commun. Process. Reprints*, pp. 442–444, Nov. 1967.

A. R. Memon, "Lowpass digital filters with linear phase," *Electron. Lett.*, vol. 6, pp. 253–254, Apr. 1970.

S. A. Miller, "A PDP-9 assembly-language program for the fast Fourier transform," Analog/Hybrid Computer Lab., Dept. Elec. Eng., Univ. Arizona, Tucson, ACL Memo. 157, Apr. 1968.

H. T. Nagle, Jr. and C. C. Carroll, "Organizing a special-purpose computer to realize digital filters for sampled-data systems," *IEEE Trans. Audio Electroacoust.*, vol. AU-16, pp. 398–413, Sept. 1968.

C. D. Negron, "Digital one-third octave spectral analysis," *J. Ass. Comput. Mach.*, vol. 13, pp. 605–614, Oct. 1966.

J. Noordanus, "Frequency synthesizers—a survey of techniques," *IEEE Trans. Commun. Technol.*, vol. COM-17, pp. 257–271, Apr. 1969.

D. J. Nowak and P. E. Schmid, "A nonrecursive digital filter for data transmission," *IEEE Trans. Audio Electroacoust.*, vol. AU-16, pp. 343–350, Sept. 1968.

A. Noyes, Jr., "Coherent decade frequency synthesizers," *Experimenter*, vol. 38, Sept. 1964.

G. C. O'Leary, "Nonrecursive digital filtering using cascade fast Fourier transformers," *IEEE Trans. Audio Electroacoust.*, vol. AU-18, pp. 177–183, June 1970.

A. V. Oppenheim, "Superposition in a class of nonlinear systems," Res. Lab. Electron., M.I.T., Cambridge, Mass., RLE Tech. Rep. 432, Mar. 31, 1965.

___, "Nonlinear filtering of convolved signals," Res. Lab. Electron., M.I.T., Cambridge, Mass., Quart. Prog. Rep. 80, pp. 168–175, Jan. 15, 1966.

___, "Realization of digital filters using block-floating-point arithmetic," *IEEE Trans. Audio Electroacoust.*, vol. AU-18, pp. 130–136, June 1970.

A. V. Oppenheim and R. W. Schafer, "Homomorphic analysis of speech," *IEEE Trans. Audio Electroacoust.*, vol. AU-16, pp. 221–226, June 1968.

A. V. Oppenheim, R. W. Schafer, and T. Stockham, "The nonlinear filtering of multiplied and convolved signals," *Proc. IEEE*, vol. 56, Aug. 1968, pp. 1264–1291.

A. V. Oppenheim and C. Weinstein, "A bound on the output of a circular convolution with application to digital filtering," *IEEE Trans. Audio Electroacoust.*, vol. AU-17, pp. 120–124, June 1969.

A. V. Oppenheim, D. Johnson, and K. Steiglitz, "Computation of spectra with unequal resolution using the FFT," *Proc. IEEE*, vol. 59, 1971, pp. 299–301.

H. J. Orchard, "The roots of the maximally flat-delay polynomials," *IEEE Trans. Circuit Theory* (Corresp.), vol. CT-12, pp. 452–454, Sept. 1965.

___, "Maximally flat approximation techniques," *Proc. IEEE*, vol. 56, Jan. 1968, pp. 65–66.

J. F. A. Ormsby, "Design of numerical filters with applications to missile data processing," *J. Ass. Comput. Mach.*, vol. 8, pp. 440–466, July 1961.

R. K. Otnes, "An elementary design procedure for digital filters," *IEEE Trans. Audio Electroacoust.*, vol. AU-16, pp. 330–336, Sept. 1964.

R. K. Otnes and L. P. McNamee, "Instability thresholds in digital filters due to coefficient rounding," *IEEE Trans. Audio Electroacoust.*, vol. AU-18, pp. 456–463, Dec. 1970.

___, "Exact second-order bandpass digital filters," *IEEE Trans. Audio Electroacoust.*, vol. AU-19, pp. 104–105 (Corresp.), Mar. 1971.

A. Papoulis, "On the approximation problem in filter design," *IRE Conv. Rec.*, part 2, pp. 175–185, 1957.

___, "Error analysis in sampling theory," *Proc. IEEE*, vol. 54, July 1966, pp. 947–955.

E. Parzen, "Notes on Fourier analysis and spectral windows," *Appl. Math. Statist. Labs.*, Stanford Univ., Stanford, Calif., Tech. Rep. 48, May 15, 1963.

___, "Statistical spectral analysis (single channel case) in 1968," Dept. Statist., Stanford Univ., Stanford, Calif., Tech. Rep. 11, ONR contract NONR-225 (80) (NR-042-234), June 10, 1968.

M. C. Pease and J. Goldberg, "Feasibility study of a special-purpose digital computer for on-line Fourier analysis," Adv. Res. Proj. Agency, Order 989, May 1967.

___, "Investigation of a special-purpose digital computer for on-line Fourier analysis," Stanford Res. Inst., Menlo Park, Calif., Special Tech. Rep. 1, Project 6557, Apr. 1967 (available from U.S. Army Missile Command, Redstone Arsenal, Ala., att: AMSMI-RNS).

M. C. Pease, "An adaptation of the fast Fourier transform for parallel processing," *J. Ass. Comput. Mach.*, vol. 15, pp. 252–264, Apr. 1968.

Y. Peless and T. Murakami, "Analysis and synthesis of transitional Butterworth-Thomson filters and bandpass amplifiers," *RCA Rev.*, vol. 18, pp. 60–94, Mar. 1957.

D. P. Petersen, "Smoothing and differential operators for digital processing of sampled-field data," *Nerem Record*, pp. 170–171, 1963.

S. E. A. Pinnell, "Design of a digital notch filter with tracking requirements," *IEEE Trans. Space Electron. Telem.* (Comment), vol. SET-10, p. 84, 1964.

M. J. Piovoso and L. P. Bolgiano, Jr., "Digital simulation using Poisson transform sequences," in *Proc. Polytechnic Inst. Brooklyn Symp. Comput. Process. Commun.*, 1969.

C. Pottle, "On the partial-fraction expansion of a rational function with multiple poles by a digital computer," *IEEE Trans. Circuit Theory*, vol. CT-11, pp. 161–162, Mar. 1964.

___, "Rapid computer time response for systems with arbitrary input signals," in *Proc. 5th Annu. Allerton Conf. Circuit System Theory*, pp. 523–533, 1967.

L. R. Rabiner, R. W. Schafer, and C. M. Rader, "The chirp z-transform algorithm and its application," *Bell Syst. Tech. J.*, vol. 48, pp. 1249–1292, May–June 1969.

___, "The chirp z-transform algorithm," *IEEE Trans. Audio Electroacoust.*, vol. AU-17, pp. 86–92, June 1969.

L. R. Rabiner, B. Gold and C. A. McGonegal, "An approach to the approximation problem for nonrecursive digital filters," *IEEE Trans. Audio Electroacoust.*, vol. AU-18, pp. 83–106, June 1970.

L. R. Rabiner and K. Steiglitz, "The design of wide-band recursive and nonrecursive digital differentiators," *IEEE Trans. Audio Electroacoust.*, vol. AU-18, pp. 204–209, June 1970.

L. R. Rabiner, "Techniques for designing finite-duration impulse-response digital filters," *IEEE Trans. Commun. Technol.*, vol. COM-19, pp. 188–195, Apr. 1971.

L. R. Rabiner and R. W. Schafer, "Recursive and nonrecursive realizations of digital filters designed by frequency sampling techniques," *IEEE Trans. Audio Electroacoust.*, vol. AU-19, pp. 200–207, Sept. 1971.

L. R. Rabiner, L. B. Jackson, R. W. Schafer, and C. H. Coker, "A hardware realization of a digital formant speech synthesizer," *IEEE Trans. Commun. Technol.*, vol. COM-19, pp. 1016–1020, Dec. 1971.

P. Rabinowitz, "Applications of linear programming to numerical analysis," *SIAM Rev.*, vol. 10, pp. 121–159, 1959.

C. M. Rader, "Speech compression simulation compiler," *J. Acoust. Soc. Am.*, vol. 37, p. 1199, June 1965.

___, "Discrete Fourier transforms when the number of data samples is prime," *Proc. IEEE*, vol. 56, June 1968, pp. 1107–1108.

___, "An improved algorithm for high-speed autocorrelation with applications to spectral estimation," *IEEE Trans. Audio Electroacoust.*, vol. AU-18, pp. 439–441, Dec. 1970.

C. Rader and B. Gold, "Digital filter design techniques in the frequency domain," *Proc. IEEE*, vol. 55, Feb. 1967, pp. 149–171.

——, "Effects of parameter quantization on the poles of a digital filter," *Proc. IEEE*, vol. 55, May 1967, pp. 688–689.

C. M. Rader, L. R. Rabiner and R. W. Schafer, "A fast method of generating digital random numbers," *Bell Syst. Tech. J.*, vol. 49, pp. 2303–2310, Nov. 1970.

C. M. Rader et al., "On digital filtering," *IEEE Trans. Audio Electroacoust.* vol. AU-16, pp. 303–315, Sept. 1968.

Q. I. Rahman, "The influence of coefficients on the zeros of polynomials," *J. London Math. Soc.*, vol. 36, part 1, pp. 57–64, Jan. 1961.

G. U. Ramos, "Roundoff error analysis of the fast Fourier transform," *Math. Comput.*, vol. 25, pp. 757–768, Oct. 1971.

R. R. Reed, "A method of computing the fast Fourier transform," M.A. thesis, Dept. Elec. Eng., Rice Univ., Houston, Tex., May 1968.

E. Renschler and B. Welling, "An integrated circuit phase-locked loop digital frequency synthesizer," Motorola Semiconductor Prod., Application Note 463.

A. A. G. Requicha and H. B. Voelcker, "Design of nonrecursive filters by specification of frequency-domain zeros," *IEEE Trans. Audio Electroacoust.*, vol. AU-18, pp. 464–471, Dec. 1970.

R. P. Rich and H. Shaw, Jr., "An application of digital filtering," *API Tech. Dig.*, vol. 4, pp. 13–18, Jan.-Feb. 1965.

J. Richalet, "Les systemes discrets," *Onde Elec.*, vol. 44, pp. 1011–1020, Oct. 1964.

R. A. Roberts and J. Tooley, "Signal processing with limited memory," in *Proc. Polytechnic Inst. Brooklyn Symp. Comput. Process. Commun.*, 1969.

H. H. Robertson, "Approximate design of digital filters," *Technometrics*, vol. 7, pp. 387–403, Aug. 1965.

E. A. Robinson and S. Treitel, "Principles of digital filtering," *Geophysics*, vol. 29, pp. 395–404, June 1964.

——, "Dispersive digital filters," *Rev. Geophys.*, vol. 3, pp. 433–461, Nov. 1965.

——, "Principles of digital Wiener filtering," Pan Am. Petroleum Corp., Tulsa, Okla., 1967.

A. E. Rogers and K. Steiglitz, "Maximum likelihood estimation of rational transfer function parameters," *IEEE Trans. Automat. Contr.*, vol. AC-12, pp. 594–597, Oct. 1967.

D. T. Ross, "Improved computational techniques for Fourier transformation," Servomechanisms Lab., M.I.T., Cambridge, Mass., Report 713-R-5, June 25, 1954.

C. Rossi, "Window functions for nonrecursive digital filters," *Electron. Lett.*, vol. 3, pp. 559–561, Dec. 1967.

E. N. Rozenvasser, "Stability criteria of nonlinear discrete systems," *Automat. Telemekh.*, vol. 27, pp. 58–66, Dec. 1966.

P. Rudnick, "Note on the calculation of Fourier series," *Math. Comput.*, vol. 20, pp. 429–430, July 1966.

M. Sablatash, "A Tellegen's theorem for digital filters," *IEEE Trans. Circuit Theory*, (Corresp.), vol. CT-18, pp. 201–203, Jan. 1971.

A. P. Sage and R. W. Burt, "Optimum design and error analysis of digital integrators for discrete system simulation," *Fall Joint Comput. Conf. AFIPS Conf. Proc.*, vol. 27. Washington, D.C.: Spartan, 1965, pp. 903–914.

A. P. Sage, "Discretization schemes and the optimal control of distributed parameter systems," in *Proc. 1st Asilomar Conf. Circuit System Theory*, 1967, pp. 191–200.

D. J. Sakrison, W. T. Ford, and J. H. Hearne, "The z-transform of a realizable time function," *IEEE Trans. Geosci. Electron.*, vol. GE-5, pp. 33–41, Sept. 1967.

I. W. Sandberg, "Floating-point-roundoff accumulation in digital filter realization," *Bell Syst. Tech. J.*, vol. 46, pp. 1775–1791, Oct. 1967.

R. W. Schafer, "Echo removal by generalized linear filtering," *Nerem Rec.*, pp. 118–119, 1967.

——, "Echo removal by discrete generalized linear filtering," Ph.D. dissertation, Dept. Elec. Eng., M.I.T., Cambridge, Mass., Feb. 1968.

R. W. Schafer and L. R. Rabiner, "Design of digital filter banks for speech analysis," *Bell Syst. Tech. J.*, vol. 50, pp. 3097–3115, Dec. 1971.

S. A. Schelkunoff, "A mathematical theory of linear arrays," *Bell Syst. Tech. J.*, vol. 22, pp. 80–107, Jan. 1943.

I. J. Schoenberg, "The finite Fourier series and elementary geometry," *Am. Math. Mon.*, vol. 57, Jun.-July 1950.

W. Schüssler and W. Winkelnkemper, "Variable digital filters," *Arch. Elek. Übertragung*, vol. 24, pp. 524–525, 1970.

W. Schüssler, "On the approximation problem in the design of digital filters," in *Proc. 5th Annu. Princeton Conf. Inform. Sci. Systems*, pp. 54–63, Mar. 1971.

A. Sekey, "A computer simulation study of real-zero interpolation," *IEEE Trans. Audio Electroacoust.*, vol. AU-18, pp. 43–54, Mar. 1970.

——, "Simulation of real-zero interpolation by the BLODIB compiler," in *Proc. Polytechnic Inst. Brooklyn Symp. Comput. Process. Commun.*, 1969.

J. L. Shanks, "Recursion filters for digital processing," *Geophysics*, vol. 23, pp. 33–51, Feb. 1967.

——, "Two planar digital filtering algorithms," in *Proc. 5th Annu. Princeton Conf. Inform. Sci. Systems*, Mar. 1971, pp. 48-53.

J. L. Shanks and T. W. Cairns, "Use of a digital convolution device to perform recursive filtering and the Cooley-Tukey algorithm," *IEEE Trans. Comput.*, vol. C-17, pp. 943–949, Oct. 1968.

J. P. Shelton, "Fast Fourier transforms and Butler matrices," *Proc. Elec. Eng.*, Australia, vol. 56, p. 350.

R. R. Shively, "A digital processor to perform a fast Fourier transform," in *Proc. 1st IEEE Comput. Conf.*, Sept. 1967, pp. 21–24.

——, "A digital processor to generate spectra in real time," *IEEE Trans. Comput.*, vol. C-17, pp. 485–491, May 1968.

R. C. Singleton, "A method for computing the fast Fourier transform with auxiliary memory and limited high-speed storage," *IEEE Trans. Audio Electroacoust.*, vol. AU-15, pp. 91–98, June 1967.

——, "On computing the fast Fourier transform," *Commun. Ass. Comput. Mach.*, vol. 10, pp. 647–654, Oct. 1967.

——, "An algol procedure for the fast Fourier transform with arbitrary factors," *Commun. Ass. Comput. Mach.*, Algorithm 339, vol. 11, pp. 776–779, Nov. 1968.

——, "Algol procedures for the fast Fourier transform," *Commun. Ass. Comput. Mach.*, Algorithm 338, vol. 11, p. 338, Nov. 1968.

——, "An algorithm for computing the mixed radix fast Fourier transform," *IEEE Trans. Audio Electroacoust.*, vol. AU-17, pp. 93–103, June 1969.

R. C. Singleton and T. C. Poulter, "Spectral analysis of the call of the male killer whale," *IEEE Trans. Audio Electroacoust.*, vol. AU-15, pp. 104–113, June 1967; also Comments by W. A. Watkins and authors' reply, vol. AU-16, p. 523, Dec. 1968.

J. B. Slaughter, "Quantization errors in digital control systems," *IEEE Trans. Automat. Contr.*, vol. AC-9, pp. 70–74, Jan. 1964.

J. M. Slazer, "Frequency analysis of digital computers operating in real time," *Proc. IRE*, vol. 42, Feb. 1954, pp. 457–466.

D. Slepian, H. O. Pollak, and H. J. Landau, "Prolate spheroidal wave functions, Fourier analysis and uncertainty," *Bell Syst. Tech. J.*, vol. 40, pp. 43–85, Jan. 1961.

E. A. Sloane, "Comparison of linearly and quadratically modified spectral estimates of Gaussian signals," *IEEE Trans. Audio Electroacoust.*, vol. AU-17, pp. 133–137, June 1969.

O. Sornmoonpin, "Investigation of quantization errors," M.S. Thesis, Univ. Manchester, England, 1966.

R. J. Stegen, "Excitation coefficients and beamwidths of Tschebyscheff arrays," *Proc. IRE*, vol. 41, Nov. 1953, pp. 1671–1674.

K. Steiglitz, "The approximation problem for digital filters," Dept. Elec. Eng., N.Y. Univ., Tech. Rep. 400-56, 1962.

——, "The general theory of digital filters with applications to spectral analysis," Ph.D. dissertation, N.Y. Univ., AFOSR rep. 64-1664, 1963.

——, "The equivalence of digital and analog signal processing," *Inform. Contr.*, vol. 8, pp. 455–467, Oct. 1965.

——, "Computer-aided design of recursive digital filters," *IEEE Trans. Audio Electroacoust.*, vol. AU-18, pp. 123–129, June 1970.

T. G. Stockham, Jr., "High-speed convolution and correlation," *1966 Spring Joint Computer Conf., AFIPS Proc.*, vol. 28. Washington, D.C.: Spartan, 1966, pp. 229–233.

——, "The application of generalized linearity to automatic gain control," *IEEE Trans. Audio Electroacoust.*, vol. AU-16, pp. 267–270, June 1968.

——, "A-D and D-A converters: their effect on digital audio fidelity," in 41st meeting Audio Eng. Soc., N.Y.C., Oct. 5–8, 1971.

J. Stoer, "A direct method for Chebyshev approximation by rational functions," *J. Ass. Comput. Mach.*, vol. 11, pp. 59–69, Jan. 1964.

D. J. Storey and R. F. Donne, "Synthesis of the Hilbert transform of a train of rectangular pulses," *Electron. Lett.*, vol. 3, pp. 126–127, Mar. 1967.

J. I. Soliman and A. Al-Shaikh, "Sampled-data controls and the bilinear transformation," *Automatica*, vol. 2, pp. 235–242, July 1965.

J. I. Soliman and H. Kwoh, "Bilinear transformation for sampled-data systems," *IEEE Trans. Automat. Contr.*, vol. AC-11, pp. 329-330, Apr. 1966.

. A. Swick, "Discrete finite Fourier transforms: a tutorial approach," Naval Res. Labs., Washington, D.C., NRL Rep. 6557, June 1967.

. C. Temes and D. A. Calahan, "Computer-aided network design—the state of the art," *Proc. IEEE*, vol. 55, Nov. 1967, pp. 1832-1863.

. S. Thelliez and J. P. Gouyet, "Introduction a l'analyse des systemes asservis a information pulsee," *Ann. Radioelec.*, vol. 16, pp. 9-68, Jan. 1961.

. Theilheimer, "A matrix version of the fast Fourier transform," *IEEE Trans. Audio Electroacoust.*, vol. AU-17, pp. 158-161, June 1969.

. J. Thomson, "Generation of Gegenbauer prewhitening filters by fast Fourier transforming," in *Proc. Polytechnic Inst. Brooklyn Symp. Comput. Process. Commun.*, 1969.

. Tierney, C. M. Rader, and B. Gold, "A digital frequency synthesizer," *IEEE Trans. Audio Electroacoust.*, vol. AU-19, pp. 48-58, Mar. 1971.

. Treitel and E. A. Robinson, "The design of high resolution digital filters," *IEEE Trans. Geosci. Electron.*, vol. GE-4, pp. 25-39, June 1966.

. Treitel, J. L. Shanks, and C. W. Frasier, "Some aspects of fan filtering," *Geophysics*, vol. 32, pp. 789-800, Oct. 1967.

. F. Trench, "A general class of discrete time-invariant filters," *J. Soc. Ind. Appl. Math.*, vol. 9, pp. 406-421, Sept. 1961.

. A. Tretter, "Some problems in the reconstruction and processing of sampled-data," Dept. Elec. Eng., Princeton Univ., Princeton, N.J., Ph.D. dissertation, pp. 105-122, Dec. 1965.

___, "Pulse-transfer-function identification using discrete orthonormal sequences," *IEEE Trans. Audio Electroacoust.*, vol. AU-18, pp. 184-187, June 1970.

. Z. Tsypkin, "Estimating the effect of quantization by level on the processes in automatic digital systems," *Automat. Telemekh.*, vol. 21, pp. 281-285, Mar. 1960.

___, "An estimate of the influence of amplitude quantization on processes in digital automatic control systems," *Automat. Telemekh.*, vol. 21, p. 195, 1960.

. W. Tufts, H. S. Hersey, and W. E. Mosier, "Effects of FFT coefficient quantization on bin frequency response," *Proc. IEEE* (Lett.), vol. 60, Jan. 1972, pp. 146-147.

M. Tsu-Han Ma, "A new mathematical approach for linear array analysis and synthesis," Ph.D. dissertation, Syracuse Univ., Syracuse, N.Y., Univ. Microfilms, 62-3040, 1961.

A. Tustin, "A method of analyzing the behavior of linear systems in terms of time series," *Proc. Inst. Elec. Eng.*, Australia, vol. 94, part IIA, May 1947, pp. 130-142.

M. L. Uhrich, "Fast Fourier transforms without sorting," *IEEE Trans. Audio Electroacoust.*, (Corresp.), vol. AU-17, pp. 170-172, June 1969.

E. Ulbrich and H. Piloty, "The design of allpass, lowpass, and bandpass filters with Chebyshev-type approximations of constant group delay," *Arch. Elek. Übertragung*, vol. 14, pp. 451-467, Oct. 1960.

H. Urkowitz, "Analysis and synthesis of delay line periodic filters," *IRE Trans. Circuit Theory*, pp. 41-53, June 1957.

R. Vich, "Selective properties of digital filters obtained by convolution approximation," *Electron. Lett.*, vol. 4, pp. 1-2, Jan. 1968.

A. J. Villasenor, "Digital spectral analysis," NASA, Washington, D.C., Tech. Note D-4510, 1968.

H. B. Voelcker, "Toward a unified theory of modulation," *Proc. IEEE*, vol. 54, pt. 1, Mar. 1966, pp. 340-353; pt. 2, May 1966, pp. 735-755.

H. B. Voelcker and E. E. Hartquist, "Digital filtering via block recursion," *IEEE Trans. Audio Electroacoust.*, vol. AU-18, pp. 169-176, June 1970.

P. J. Walsh, "A study of digital filters," Tech. Rep. Ad 71-0381, Dec. 1969.

W. Wasow, "Discrete approximations to the Laplace transformation,"

Z. Angew. Math. Phys., vol. 8, pp. 401-417, 1957.

D. G. Watts, "A general theory of amplitude quantization with applications to correlation determination," *Proc. Inst. Elec. Eng.*, Australia, vol. 109C, 1962, p. 209.

___, "Optimal windows for power spectra estimation," Math. Res. Center, Univ. Wisconsin, MRC-TCR-506, Sept. 1964.

C. S. Weaver, P. E. Mantey, R. W. Lawrence, and C. A. Cole, "Digital spectrum analyzers," Stanford Electron. Labs., Stanford, Calif., Rep. SEL 66-059 (Tr-109-1/1810-1), June 1966.

C. S. Weaver et al., "Digital filtering with applications to electrocardiogram processing," *IEEE Trans. Audio Electroacoust.*, vol. AU-16, pp. 350-392, Sept. 1968.

C. J. Weinstein, "Quantization effects in digital filters," Lincoln Lab. Tech. Rep. 468, Nov. 21, 1969.

C. Weinstein, "Quantization effects in frequency sampling filters," *Nerem Record*, p. 222, 1968.

___, "Quantization effects in digital filters," Ph.D. dissertation, Dept. Elec. Eng., M.I.T., Cambridge, Mass., July 1969.

___, "Roundoff noise in floating point fast Fourier transform computation," *IEEE Trans. Audio Electroacoust.*, vol. AU-17, pp. 209-215, Sept. 1969.

C. Weinstein and A. V. Oppenheim, "A comparison of roundoff noise in floating-point and fixed point digital filter realizations," *Proc. IEEE* (Lett.), vol. 57, June 1969, pp. 1181-1183.

P. D. Welch, "A direct digital method of power spectrum estimation," *IBM J. Res. Develop.*, vol. 5, pp. 141-156, Apr. 1961.

___, "The use of the FFT for estimation of power spectra: a method based on averaging over short, modified periodograms," *IEEE Trans. Audio Electroacoust.*, vol. AU-15, pp. 70-73, June 1967.

___, "A fixed-point fast Fourier transform error analysis," *IEEE Trans. Audio Electroacoust.*, vol. AU-17, pp. 151-157, June 1969.

M. A. Wesley, "Associative parallel processing for the fast Fourier transform," *IEEE Trans. Audio Electroacoust.*, vol. AU-17, pp. 162-165, June 1969.

J. E. Whelchel, Jr. and D. F. Guinn, "The fast Fourier-Hadamard transform and its use in signal representation and classification," *1968 Eascon Conv. Rec.*, pp. 561-571.

___, "FFT organizations for high-speed digital filtering," *IEEE Trans. Audio Electroacoust.*, vol. AU-18, pp. 159-168, June 1970.

W. D. White and A. E. Ruvin, "Recent advances in the synthesis of comb filters," *IRE Nat. Conv. Rec.*, pp. 186-199, 1957.

D. E. Whitney, "Computation errors in inertial navigation systems which employ digital computers," Instrum. Lab., Cambridge, Mass., T-409, Feb. 1965.

J. R. B. Whittlesey, "A rapid method for digital filtering," *Commun. Ass. Comput. Mach.*, vol. 7, pp. 552-556, Sept. 1964.

B. Widrow, "A study of rough amplitude quantization by means of Nyquist sampling theory," *IRE Trans. Circuit Theory*, vol. CT-3, pp. 266-276, Dec. 1956.

___, "Statistical analysis of amplitude-quantized sampled-data systems," *AIEE Trans.* (Appl. Ind.), vol. 79, part 2, pp. 555-568, 1961.

W. Winkelnkemper, "Unsymmetrical bandpass and bandstop digital filters," *Electron. Lett.*, vol. 5, Nov. 13, 1969.

J. H. Wilkinson, "Error analysis of floating-point comparison," *Numer. Math.*, vol. 2, pp. 319-340, 1960.

J. C. Wilson, "Computer calculation of discrete Fourier transforms using the fast Fourier transform," Center Naval Analyses, Arlington, Va., OEG Res. Contrib. 81, June 1968.

T. Y. Young, "Representation and analysis of signals, part X. Signal theory and electrocardiography," Dept. Elec. Eng., Johns Hopkins Univ., Baltimore, Md., May 1962.

___, "Binomial-weighted orthogonal polynomials," *J. Ass. Comput. Mach.*, vol. 14, pp. 120-127, Jan. 1967.

A. I. Zverev, "Digital MTI radar filters," *IEEE Trans. Audio Electroacoust.*, vol. AU-16, pp. 422-432, Sept. 1968.

IEEE

COMPUTER SOCIETY

Press Activities Board

IEEE Computer Society Publications

The world-renowned Computer Society publishes, promotes, and distributes a wide variety of authoritative computer science and engineering texts. These books are available in two formats: 100 percent original material by authors preeminent in their field who focus on relevant topics and cutting-edge research, and reprint collections consisting of carefully selected groups of previously published papers with accompanying original introductory and explanatory text.

Submission of proposals: For guidelines and information on Computer Society books, send e-mail to cs.books@computer.org or write to the Project Editor, IEEE Computer Society, P.O. Box 3014, 10662 Los Vaqueros Circle, Los Alamitos, CA 90720-1314. Telephone +1 714-821-8380. FAX +1 714-761-1784.

IEEE Computer Society Proceedings

The Computer Society also produces and actively promotes the proceedings of more than 130 acclaimed international conferences each year in multimedia formats that include hard and softcover books, CD-ROMs, videos, and on-line publications.

For information on Computer Society proceedings, send e-mail to cs.books@computer.org or write to Proceedings, IEEE Computer Society, P.O. Box 3014, 10662 Los Vaqueros Circle, Los Alamitos, CA 90720-1314. Telephone +1 714-821-8380. FAX +1 714-761-1784.

Additional information regarding the Computer Society, conferences and proceedings, CD-ROMs, videos, and books can also be accessed from our web site at http://computer.org/cspress

4/16/98